Introducing Groundwater

Michael Price

The University of Reading, UK
and
British Geological Survey

With a Foreword by
William Back
US Geological Survey

CHAPMAN & HALL

London · Weinheim · New York · Tokyo · Melbourne · Madras

Published by Chapman & Hall, 2–6 Boundary Row, London SE1 8HN, UK

Chapman & Hall, 2–6 Boundary Row, London SE1 8HN, UK

Chapman & Hall GmbH, Pappelallee 3, 69469 Weinheim, Germany

Chapman & Hall USA, 115 Fifth Avenue, New York, NY 10003, USA

Chapman & Hall Japan, ITP-Japan, Kyowa Building, 3F, 2–2–1 Hirakawacho, Chiyoda-ku, Tokyo 102, Japan

Chapman & Hall Australia, 102 Dodds Street, South Melbourne, Victoria 3205, Australia

Chapman & Hall India, R. Seshadri, 32 Second Main Road, CIT East, Madras 600 035, India

First edition 1985
Reprinted 1991, 1992, 1994
Second edition 1996
Reprinted 1996, 1997

© 1996 Michael Price

Typeset in 10/12pt Times by WestKey Ltd, Falmouth, Cornwall
Printed in Great Britain at St Edmundsbury Press, Bury St Edmunds, Suffolk

ISBN 0 412 48500 1

A catalogue record for this book is available from the British Library

Library of Congress Catalog Card Number: 95-69070

∞ Printed on permanent acid-free text paper, manufactured in accordance with ANSI/NISO Z39.48–1992 and ANSI/NISO Z39.48–1984 (Permanence of Paper).

Introducing Groundwater

This book is dedicated to the world's children

Contents

Foreword

You are about to embark on a stimulating intellectual journey touching on many fields of science as you begin the study of groundwater. You are fortunate to have such an understandable book that explains the complex concepts in such a manner that the ideas can be quickly grasped and long retained. You will soon understand answers to dozens of questions, such as: Where does the water in a well come from? Does underground water move like a river or is it more quiet like a lake? Is a drink from a well always cold? Is groundwater more pure than river water? How do springs form? Were there more springs in historic times than there are now, if so why? How do caves form? Why aren't they full of water? And the list goes on and on. The answers to such questions have been deduced from the application of principles from chemistry, physics, biology, geomorphology and other earth sciences. Now more than ever, it is important for the non-scientist to understand the principles of hydrogeology. Groundwater is such an important resource, and the behaviour and quality of groundwater – including changes caused by contamination – can have such profound effects on the environment, that an understanding of it is a requisite for rational decisions concerning all resource management. A democracy requires that an informed public be able to identify options, choose alternatives and understand the consequences of its actions or non-action. Effective environmental regulation must be based on an understanding of groundwater and other geosciences in order to identify those goals that are desirable and achievable.

Students who continue their studies to become professional groundwater scientists or engineers will have a most rewarding career in meeting the scientific and technical challenges that must be overcome in order to sustain a viable environment. Throughout the world they work for (1) government regulatory and research agencies, (2) environmental and engineering consulting companies, (3) industries, and (4) as teachers and researchers at universities. The basic goal of the work of all hydrogeologists is to maintain, or to improve, the health of the public and the environment. The science of groundwater is the fastest growing discipline within the earth sciences. This is a reflection of the growing awareness of the value and fragility of water resources, and the response and desire of government, the public and industry to provide enlightened management of resources. Along with this recognition of environmental problems is recognition of the need for research in hydrogeology by universities and other research institutions.

Readers are most fortunate to have a clearly written book by an eminent hydrogeologist, whose broad experience permits him to explain the principles and demonstrate their consequences. You will enjoy reading and studying this book because of the way the author develops the concepts, explains the terminology and provides essential information without extraneous detail that could distract from the basic understanding. You will resolve the meaning and use of terms such as transmissivity, storage coefficient, specific yield and others you may have seen in literature on groundwater. As Professor Ken Howard, University of Toronto, stated in a review of the first edition 'Michael Price's new book is more than just an introduction to groundwater. It is a lesson in expressive and informative writing that we would all do well to heed.' This statement is equally true for this edition, and I know that the hours spent with the book will be most rewarding.

William Back
Senior Research Scientist
US Geological Survey
Reston, Virginia, USA

Preface

It is nearly ten years since the first edition of this book appeared. During that period knowledge has advanced, and concern for the environment has increased. There have been major droughts in Britain and other parts of Europe, in the United States, Africa and Australia, and increasing suggestions that the climate is being changed by human activities. More people have become concerned about possible long-term shortages of water, and about the effects that trying to supply human needs for water might have on the environment. There is increasing concern about pollution. More complex legislation has been introduced to try to cope with some of these problems.

In this new edition I have tried to cover these topics as they relate to hydrogeology. The chapter on water quality has been enlarged to include more information on chemistry and micro-organisms and a new chapter (Chapter 13) has been added specifically on pollution. At the same time the rest of the book has been revised to bring it up to date, and I have included more mention of legislation. In making these changes I have tried to retain the essential quality of the first edition, which – I hope – was its accessibility to a non-specialist. To this end I have increased the number of boxes within the text that deal with some of the more difficult – if more interesting – concepts.

Michael Price
Reading, January 1995

Preface to the first edition

Although there are several excellent books on groundwater, it remains for many people a misunderstood and even mysterious substance. It seems to me that this is because all the textbooks on the subject are aimed at people who are, or intend to become, specialists in the subject; in effect the books are preaching to the converted. On the other hand, many of the more attractive introductory books on water mention groundwater only briefly. Whatever the reason, there is an unfortunate lack of understanding of the subject, even among geologists, geographers and civil engineers.

This book is an attempt to provide the non-specialist with a readable introduction to the subject. I have kept technical terms to a minimum, and where I have used them I have tried to explain why they are used. The emphasis is on principles rather than detail.

In a book of this kind little or nothing is original. However, I have assumed that readers will not want to see every factual statement supported by a reference; I have also assumed that anyone who wants more information will be likely to proceed initially to a more advanced book rather than to papers in academic journals. Therefore I have given references to books, and have referred only to those papers which are too recent to have been incorporated into books, or which are 'classics', worth looking at for that reason, and to those whose content has escaped mention in the existing books. To avoid breaking up the text, I have simply listed selected references at the end of each chapter.

The aim of the book is simplicity throughout, and I have kept mathematical formulae to a minimum. The entire text should be within the scope of anyone who has studied mathematics and physics or chemistry to GCE 'O' Level standard or equivalent. A knowledge of geology, though helpful, is not essential. I have relegated some of the more difficult concepts to self-contained boxes, which can be bypassed on first reading or by those who wish to avoid detail.

Michael Price

Acknowledgements

Many people have contributed to this book by supplying help and advice. I am particularly grateful to my friends and colleagues at the British Geological Survey and in universities, colleges and water authorities in the United Kingdom, who have not only helped directly in the writing of the book but who have added so much to my education over the years. During part of the time that I was working on the manuscript, I was fortunate enough to spend a year with the Geological Survey of Canada, and I should like to acknowledge the help and advice I received from friends at the GSC and elsewhere in Canada.

Specifically, I should like to thank John Barker for checking the accuracy of the physical statements, Adrian Bath for advising on water chemistry and Professor K.J. Ives for help with the public health aspects of water quality. Data and information on specialist topics were supplied by Southern Water Authority, Failing Supply Limited, and by Sigmund Pulsometer Pumps; I am particularly grateful to Graham Daw and Cementation Specialist Holdings Limited for advice on geotechnical problems. My special thanks are due to Dick Downing, whose seemingly inexhaustible knowledge of hydrogeology has been called upon throughout.

Early parts of the manuscript were reviewed by B.P.J. Williams and M.S. Money who provided useful criticism, and the entire manuscript has been reviewed by Dick Downing, Chris Wilson and Brian Knapp. Any faults that remain are the results of my stubbornness, not their omission. I am also grateful to Roger Jones and his colleagues at George Allen and Unwin for guiding the work to completion.

Finally, I am indebted to Mary, Susan and Ian for their patience and tolerance during my prolonged mental absences from family life!

ADDITIONAL ACKNOWLEDGEMENTS TO THE SECOND EDITION

In addition to those people and organizations listed above, many people have contributed to the preparation of this revised edition. I am grateful to all the people who reviewed the book, or sent in comments, for their constructive criticism. I should like to thank my colleagues at the British Geological Survey, the Institute of Hydrology, the University of Reading, and friends elsewhere in the water industry, for continuing the process of my education. I am especially grateful to Lester Simmonds of the Department of Soil Science at Reading for critically reviewing Chapters 4 and 5; to Martin Wood of the same department and Julie West of BGS for help with microbiology; and to John Chilton (BGS) and Kevin Wilson (Ready Mixed Concrete (UK) Ltd) for help with Chapter 13. Once again, any faults that remain arise from my failure to accept their advice, not from their failure to see them.

Dominic Recaldin and Una-Jane Winfield at Chapman & Hall encouraged, wheedled and coerced me to keep going, and were patient with the delays caused by my move to Reading.

As before, I am indebted to Mary, Sue and Ian for their forbearance.

Michael Price

Introduction

Every year about 110 million million cubic metres (m^3) of water falls as precipitation on the land areas of the Earth. With a world population estimated at 5700 million, that means that we each receive on average about 53 000 litres, or 53 tonnes, of fresh water per day, and that is without considering the water that falls as rain and snow over the oceans. Surely, there should be enough water to go around.

Yet in spite of this apparent abundance, thousands of people die every day from diseases associated with inadequate supplies of clean water. The problem of course is not the amount of water, nor the number of people, but the relative distribution of the two – and the fact that the water arrives unevenly over time.

To counter these problems, people have always tried to place their settlements near reliable sources of water, where there is some natural feature that somehow smoothes out the uneven distribution of the rainfall – a lake to store it, perhaps, or a river bringing water from a region with a heavier or more reliable rainfall. It would be difficult to name any major town in the United Kingdom, or a capital city anywhere in the world, that does not lie on the banks of a river. There are admittedly other reasons for this – defence and communication being obvious ones – but the availability of water must always have been a key factor.

There are of course some areas of the world where, for reasons of climate or of geology, there are no permanent streams or rivers; yet many of these areas have been settled for thousands of years. The inhabitants of the early settlements relied for their supplies on water which occurs underground, often within a few metres of the surface, and which they exploited by digging wells. Sometimes the names of the settlements – names that in Britain end in '-well' or in the Middle East begin with 'Bir' or 'Beer' – testify to the nature of their water supply.

In the early cultures of Britain and the Middle East the origin of underground water may not have been understood, but its existence was known and exploited. In Britain today many people not only do not realize that much of the nation's water supply is drawn from underground, they are also unaware that underground water is common and widespread. As most people in Britain take their water supplies, along with their energy supplies, for granted, it is unlikely that they would stop to wonder where the water was coming from until the day when it failed to arrive; 'you never miss the water till the well runs dry' is a saying with a literal meaning in addition to the figurative one. The lack of awareness is perhaps inevitable, because the underground supplies are invisible and so cannot have the same impact on the senses as the expanse of a lake or the roar of a waterfall. It is however unfortunate, because underground supplies represent the largest accessible store of fresh water on Earth. They also frequently provide the best – in some cases the only – solution to the problem of providing water for drinking and irrigation in the developing world.

How does this underground water occur? How widespread is it, and how reliable as a source of supply? How can it be protected and best utilized for the benefit of people? A science called **hydrogeology** – a branch of geology devoted to the study of underground water – has grown up in an attempt to provide detailed answers to these and related questions. This book is an attempt to provide some of the answers in general terms for those people who are curious about one of our most precious assets – people who want to think about the water *before* the well runs dry.

Water underground

In Xanadu did Kubla Khan
A stately pleasure-dome decree:
Where Alph, the sacred river, ran
Through caverns measureless to man
Down to a sunless sea.

Samuel Taylor Coleridge

Many people share with Coleridge the idea that underground water occurs in vast lakes in caverns. They picture the water as flowing from one lake to another along underground rivers. Successful wells or boreholes, they imagine, are those that intersect these lakes or rivers; unsuccessful ones are those that encounter only 'solid' rock. The art of the water diviner or dowser is seen as predicting the location of these postulated subterranean watercourses, and so selecting a site for a borehole where water will be struck.

These popular misconceptions probably grew up because the only places where it is possible to see underground water in its natural state are the spectacular caverns that occur in hard limestones such as those of the Mendips and the Peak District in Britain, and in Kentucky and New Mexico in the USA. Visitors to these and similar caves are told that they were formed by the action of underground water. They hear of rivers that disappear into the ground in one place to reappear elsewhere, having flowed underground for parts of their courses. Not surprisingly, many of the visitors form the impression that this must be the normal mode of occurrence of all subsurface water. If this were so, since people are never slow to make money from others, we might expect to find many other places where tourists could venture into the ground to see such features. The fact that they are relatively rare suggests either that underground water is restricted in its occurrence, or that it normally occupies less spectacular habitats. The latter possibility is borne out by the fact that in the regions where caverns are common, wells and boreholes are rare. Those parts of Britain that draw heavily on underground water for their supplies – such as the chalk lands

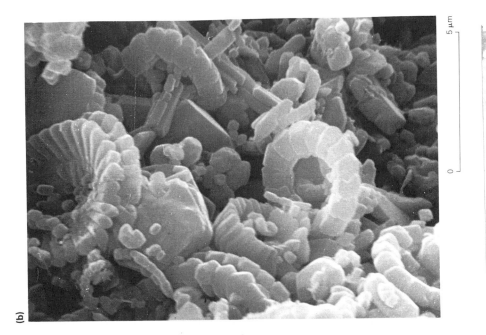

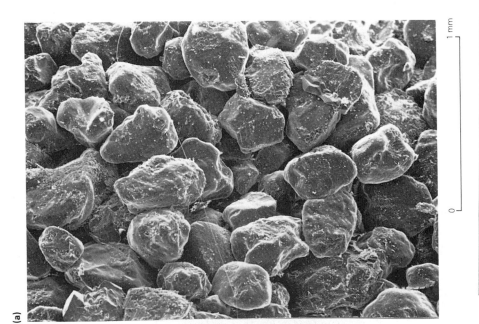

Figure 2.1 Different types and sizes of voids in rock. (a) Photograph taken with an electron microscope (an electron photomicrograph) of sandstone of Permian age from a borehole in Cumbria, England. The porosity is 31%. (b) Electron photomicrograph of chalk from a borehole in Berkshire, England, showing the minute fossils and shell fragments that make up the chalk. The porosity is 46% but the small size of the pores means that in the absence of fissures the permeability is low. (c) Pore space resulting from fissures in limestone of Jurassic age (Great Oolite) in a quarry in the Cotswold Hills, England. Such fissures contribute most of the permeability of this aquifer. (d) A fissure which has been greatly enlarged by solution in Jurassic (Inferior Oolite) limestone, Cotswold Hills. ((a) and (b) are reproduced by permission of the Director, British Geological Survey.)

of southern England and the sandstone tracts of the Midlands – contain few natural caves.

To understand the ways in which water occurs underground, we first need to think a little about the ground itself. Britain, like the rest of the Earth's crust, is made up of rocks of various types; in many parts of Britain these rocks are not readily seen, because they are hidden by soil and the landscape is clothed by vegetation. But rocks can be seen in seaside cliffs, on mountainsides and, among other places, in quarries and cuttings. Wherever we live, if we probe deeply enough below the surface of the Earth, we shall eventually encounter rock.

In Chapter 7 we shall look in more detail at different types of rock, but for the present we need to note only one feature. This is that nearly all rocks in the upper part of the Earth's crust, whatever their type, age or origin, contain openings called **pores** or **voids**. These voids (Figure 2.1) come in all shapes and sizes. Some of them – like the tiny pores in the chalk of the English downlands – are too small to be seen with the unaided eye. However, in exceptional cases, the openings may be tens of metres across, like the limestone caverns referred to earlier.

One common rock, sandstone, has pores that are more easily visualized (Figure 2.1a). If you take a handful of sand and look at it closely – preferably through a magnifying glass – you will see that there are numerous tiny openings between the grains of sand. If you cannot readily find a handful of sand, look at granulated sugar which shows exactly the same feature. Sandstone is merely sand that has turned into rock because the grains of sand have become cemented together, and most of the openings will usually have been retained in the process – it is as though our granulated sugar had become a block of sugar.

The property of a rock of possessing pores or voids is called **porosity**. Rocks containing a relatively large proportion of void space are described as 'porous' or said to possess 'high porosity'. Soil also is porous. On a hot summer day the surface soil may appear quite dry, but if we dig down a little way the soil feels damp; if we could dig far enough to reach rock, this too would feel damp. The reason for this is that the pores are not all empty; some of them are filled, or partly filled, with water. In general it is the smaller pores that are full and the larger ones that are empty. At a still greater depth we should find that all the pores are completely filled with water, and we should describe the rock or soil as 'saturated'. In scientific terms we should have passed from the **unsaturated zone** to the **saturated zone**.

If we dig or drill a hole from the ground surface down into the saturated zone, water will flow from the rock into our hole until the water reaches a constant level. This will usually be at about the level below which all the pores in the rock are filled with water – in other words, the upper limit of the saturated zone. We call this level the **water table**.

The distance we need to drill or dig to reach the water table varies from place to place; it may be less than a metre, or more than a hundred metres.

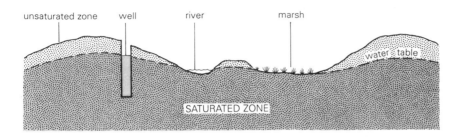

Figure 2.2 The water table.

In general the water table is not flat; it rises and falls with the ground surface but in a subdued way, so that it is deeper beneath hills and shallower beneath valleys (Figure 2.2). It may even coincide with the ground surface. If it does we can easily tell, because the ground will be wet and marshy or there will be a pond, spring or river. Where the water table is below the ground, as is usual, its depth can be measured in a well.

All water that occurs naturally below the Earth's surface is called **subsurface water**, whether it occurs in the saturated or unsaturated zones. (I insert the word 'naturally' so as to exclude from the definition water in pipes, and the like.) Water in the saturated zone, that is to say below the water table, is called ground water or **groundwater**. Early practice was to write the term as two words, but in Britain – and more recently in North America – there is an increasing tendency to write it as one word, to emphasize the fact that it is a technical term with a particular meaning.

Why is groundwater important? One reason, mentioned in the introduction, is that in some areas it is the only source of water. A second reason is that groundwater represents a major proportion of the Earth's usable water resources.

It is estimated that the total amount of water on the Earth is a little over 1400 million cubic kilometres (km^3). Of this total about 1370 million km^3, or about 95%, is sea water. Much of the remainder – about 2% of the total – occurs in solid form in glaciers and polar ice-caps. Virtually all of the remaining water – the non-marine, unfrozen water – is groundwater. The water in rivers and lakes, in the atmosphere and in the unsaturated zone, together amounts to only about 1/50th of 1% of the world's total water supply.

For reasons that we shall see later, it is probably more difficult to make an accurate estimate of the volume of groundwater on the Earth than it is to estimate any other component of the total water resources of the planet. Estimates of the volume of groundwater range from around 7 million km^3 to as much as 60 million km^3. Not all of it is usable: some is held in pore spaces that are so small, or in rocks that are so deep, that it is not really a resource. Similarly, some occurs in permanently frozen ground in high latitudes, and more is saline. Nor is groundwater distributed uniformly over the land areas of the world. Where it is accessible, it can be used – and has been used – literally to make deserts bloom, to make the difference between wilderness and plenty.

In terms simply of quantity, groundwater is therefore of great import-ance. But quantity is not everything, and groundwater has several other advantages over surface water as a source of supply. A surface reservoir must usually be impounded at one time, in one place, even though its full capacity may not be needed for many years. Groundwater can often be developed where and when it is needed, by sinking boreholes one at a time in appropriate places; if demand for water increases less rapidly than expected, the water-supply authority is not left with an expensive liability in the form of a man-made lake that nobody needs.

Unlike surface-water reservoirs which occupy large areas, frequently of prime agricultural land, the presence and utilization of groundwater need not conflict with other use of the land under which it occurs. Deep beneath the ground it is unseen, insulated from changes in temperature, and pro-tected from evaporation which, in a hot summer, can cause substantial losses of water from reservoirs and lakes. Its depth also renders ground-water less vulnerable to pollution, which is a potential threat to water on the surface.

The pollution risk to surface water is exemplified by an incident which occurred on 6 April 1978 in Yorkshire, England, when a barn containing stores of herbicides and other pesticides caught fire. The chemicals, released when their containers were burnt or melted in the heat, were washed into drainage ditches by the water used in fire-fighting. The substances included a paraquat-based weedkiller, which in high concentrations is a deadly poison to which there is no known antidote. The contaminated water travelled along the ditches into the River Kyle, a tributary of the River Ouse, and so into the Ouse itself. The city of York draws most of its water supply from the Ouse.

On hearing the news of the incident, the York Water Works Company prudently closed its river intakes until the danger had passed. The Company's distribution reservoirs were sufficient to maintain the supply, but some consumers were inconvenienced as a result of the reduction in mains pressure. Had the danger not been immediately understood, or had the authorities been slower to respond, the consequences could have been serious.

Such incidents continue to occur. In July 1992 the rivers Aire and Calder, by coincidence also in Yorkshire, were reported polluted after another fire at a chemical store. Although it would be foolish to imagine that ground-water is immune from contamination, its natural cover of soil and rock provides it with considerable protection.

If groundwater has such advantages – abundance, availability in arid climates, and relative safety from pollution – one may be forgiven for wondering why we do not use it exclusively, or why we sacrifice large tracts of land in areas of outstanding natural beauty as reservoir sites. Surely there must be some drawbacks? There are.

Groundwater, despite its many advantages, has one serious disadvantage – it is not uniformly distributed throughout the Earth's crust. There are large areas of Britain, and of the world, where groundwater cannot be obtained in sufficient quantities to justify the expense of sinking wells and boreholes. At first sight this may seem to contradict what was said earlier, that nearly all rocks are to some extent porous and that below a certain depth – the water table – the pores are filled with groundwater. Since rocks are ubiquitous, it might seem reasonable to conclude that groundwater is available everywhere. It might seem reasonable, but it would be wrong.

Three main factors complicate the issue. The first is the extent to which the rocks are porous. If there are only a few small voids then the amount of water contained in a given volume of rock will be limited. So that we can make quantitative comparisons between different rocks, we define **porosity** as the ratio of the volume of the voids in the rock to the total volume of the rock. We usually express this ratio as a percentage, and so we speak for example of rocks with 20% or 30% porosity, meaning that the voids occupy respectively 20 or 30% of the total rock volume.

In algebraic notation,

$$n = \frac{V_p}{V_b} \qquad\qquad (2.1)$$

or

$$n = \frac{V_p}{V_b} \times 100\%$$

where n = porosity, V_p = volume of voids, and V_b = bulk volume of rock.

The second factor is a combination of the size of the pores and the degree to which the pores are interconnected, because this combination will control the ease with which water can flow through the rock. We call this factor the **permeability**. Materials that allow water to pass through them easily are said to be **permeable**; those that permit water to pass only with difficulty, or not at all, are **impermeable**. A rock may be porous but relatively impermeable, either because the pores are not connected or because they are so small that water can be forced through them only with difficulty (Figure 2.1b). Conversely, a rock that has no voids except for one or two open cracks will have a low porosity, and will be a poor store of water, but because water will be able to pass easily through the cracks the permeability will be high. Layers of rock sufficiently porous to store water *and* permeable enough to allow water to flow through them in economic quantities are called **aquifers**.

At great depths – typically 10 km or so below the Earth's surface – the rocks are believed to be so compressed and altered as a result of their deep burial that the voids have been closed and, for practical purposes, all the rocks are assumed to be impermeable. An interesting demonstration of this was provided by the Kola 'Superdeep' – the 12 km deep exploratory borehole that Soviet scientists began drilling in the 1970s into the rocks of the Kola Peninsula in Arctic Russia. Water entered the well in limited quantities through cracks (fissures) in the rock throughout most of its depth; however, below 9000 m (9 km) the flows lasted only for a short time, suggesting that the fissures were not continuous. Above 9 km the inflows appeared to continue indefinitely. These deep, impermeable rocks form a floor below which groundwater cannot move – in effect they represent the bottom of the groundwater storage space available in the Earth's crust. Above this floor the storage space is filled, up to the level of the water table, with water that has entered the rocks – usually from rainfall.

If we knew the porosity of all the various rock formations that make up the Earth's crust, and the depths to which they are porous, then we should be able to make a reasonably accurate estimate of the volume of groundwater on Earth. Since we do not know these things, the calculations are rather imprecise. Essentially, what we do is to take the total area of land (149 million km^2) and multiply it by the depth to which we expect porous rock to be present. This gives us the volume of rocks that are likely to contain groundwater. We then multiply this volume by an average porosity (expressed as a fraction) to arrive at a value for the quantity of groundwater in storage. We can refine this value by excluding areas like Antarctica, where the rocks are mostly covered by large thicknesses of ice, and other areas where the groundwater is likely to be frozen and therefore not available.

One of the first people to attempt the calculation with any accuracy was Raymond Nace of the United States Geological Survey. Excluding Antarctica and permanently frozen regions, he calculated a volume of fresh groundwater in storage of 7 million km^3; this was based on the assumption of an average porosity of 5% to a depth of 1000 m. Nace recognized that there is probably at least that volume of water held in rocks below 1000 m but concluded, probably correctly, that much of it is probably saline. His calculation excludes water held in the pores of rocks such as clays and shales; these rocks are porous but not permeable, so that the groundwater that they contain is not readily available as a resource.

In some areas, rocks of low permeability (such as clays or shales) may occur at or near the ground surface. None of these rocks are totally impermeable, but if they occur at the surface they will usually restrict the amount of rainfall that can soak into the ground. Similarly, if one of these relatively impermeable layers occurs *beneath* an aquifer, it will restrict the movement of water downwards from the aquifer. If part of the aquifer is also overlain by one of these layers, as in Figure 2.3, movement of water from the aquifer becomes so restricted that the groundwater in the aquifer becomes **confined** under pressure. In Figure 2.3, for example, rain water can enter the aquifer between X and Y. As a result of this supply of water the

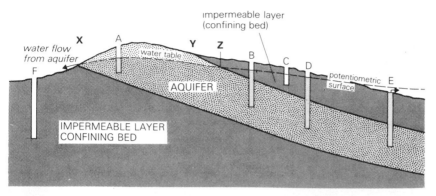

Figure 2.3 A confined aquifer and its potentiometric surface. Between X and Z the aquifer is unconfined and has a water table; to the right of Z the aquifer is confined and has a potentiometric surface. Wells A, B, D and E enter permeable material and strike water. Wells C and F are in impermeable material which will yield water only very slowly.

storage in the aquifer 'fills up' to give the water table in the position shown; the water table has to be above X, since this is the only place where water can flow from the aquifer. To the right of Z, the groundwater is confined between the impermeable layers.

The layers of low permeability are called **confining layers** or **confining beds**: the aquifer is called a **confined aquifer**. It is important to note, however, that between X and Z the aquifer is **unconfined**. To the right of Z, where the aquifer is confined, it has no water table and no unsaturated zone – the permeable material is saturated for its full thickness. If a well is dug or drilled into the upper confining layer (for example, well C in Figure 2.3), it will encounter no groundwater (except perhaps a very slow seepage from the 'impermeable' material); if the well is deep enough to reach the aquifer (as at B and D in Figure 2.3) the water will rise up in the well because the water in the aquifer is under pressure. The level at which the water stands in the wells defines an imaginary surface whose height above the aquifer depends on the pressure in the aquifer; this surface is called the **potentio-metric surface**. Sometimes the potentiometric surface may rise above ground level, in which case the well (like E in Figure 2.3) will overflow.

We shall talk more about confined and unconfined aquifers in Chapter 7, but for now we have seen something of the influence of permeability on the availability of groundwater.

The third factor that determines the amount of groundwater available from the rocks of an area is the amount of replenishment – the degree to which water abstracted from the aquifer is replaced. The replenishment may come from above, as a result of rainfall soaking into the ground, or it may take place laterally or from below, from adjacent aquifers carrying water from elsewhere. The replenishment factor depends not only on the nature of the rocks but on the soil and vegetation that cover them and on the climate of the region. It is part of the **water balance** of the area – the balance between the water that enters the area and the water that is used or which leaves it. In assessing the groundwater resources of any region, knowledge of the water balance is as vital as knowledge of the porosity and permeability of the rocks. This is because groundwater is not isolated from other water; as we have seen, it is part of the Earth's total store of water. As such, it is in more or less continuous interchange with all other water in a system of circulation called the **water cycle**, which is the subject of the next chapter.

SELECTED REFERENCES

Kozlovsky, Ye.A. 1987. *The superdeep well of the Kola Peninsula*. Berlin: Springer-Verlag.

Nace, R.L. 1969. World water inventory and control. Chapter 2 of *Water, earth and man* (ed. R.J. Chorley). London: Methuen.

UNESCO 1971. Scientific framework of world water balance. *Technical papers in hydrology*. Paris: UNESCO.

Water in circulation | 3

It is a matter of observation that virtually all rivers discharge their waters into the sea. As the level of the sea remains more or less constant, and as the rivers show no apparent sign of ceasing to flow, there must be some mechanism by which water is returned from the sea to the land at the same average rate at which it flows via the rivers to the sea. The nature of this return mechanism is less obvious than the clearly visible flow of the rivers, and has occasioned much speculation throughout history.

We now know that the return path is through the atmosphere. The energy of the Sun's rays causes water to evaporate from the surface of the oceans (Figure 3.1). The water vapour so produced is a normal part of the Earth's atmosphere; it remains in the atmosphere, completely invisible, unless cooled sufficiently to cause it to condense and form water droplets. The cooling occurs as a result of the vapour-laden air rising into higher and colder regions of the atmosphere. This may result from the air being forced upwards over a physical barrier, such as a mountain range, or over a meteorological barrier, such as a mass of colder, denser air; or it may result from convection. Whatever the cause, the chilled vapour condenses as droplets which form around suitable nuclei (tiny dust and other particles) that are nearly always present in the atmosphere. When the droplets are present in sufficient quantity, we see them as clouds.

These water droplets are much smaller than raindrops, and are light enough to remain suspended in the atmosphere. The process by which they grow until they are large enough to fall as rain is surprisingly complicated. To me, living in a temperate, humid region, where rain is generally regarded as a persistent nuisance until we have been without it for a few weeks, it came as something of a shock to learn that it actually needs a favourable combination of circumstances to produce rain, but such is the case.

A typical thundercloud contains more than 50 000 tonnes of water. At any one time, the atmosphere contains about 13 000 km^3 or 13×10^{12} tonnes of water. These may sound enormous amounts, but the total atmospheric water is in reality about 1/1000th of 1% of the Earth's water, or 0.03% of non-oceanic water. If all this atmospheric water were simultaneously to fall

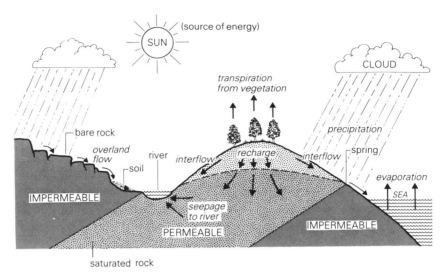

Figure 3.1 Diagrammatic representation of the water cycle.

as rain, it would produce an average of about 25 mm of rainfall over the total surface of the Earth. The average annual precipitation over the Earth is believed to be about 1000 mm, which means that the equivalent of about 40 total condensations of the atmospheric water occur each year.

The water cycle is usually depicted as a simple system of circulation that takes water from the sea as vapour, deposits it on land as precipitation, and returns it quickly to the sea by way of rivers. In this system, the ocean is the only large store of water, and the only place where a water molecule resides for any length of time. In reality, the cycle is transferring water between stores (Figure 3.2). The ocean is the largest store, but water molecules may spend long periods of time in ice-caps or in the pore spaces of rocks. We can see from Figure 3.2 that the atmosphere and the rivers may hold relatively tiny amounts of the Earth's water resources, but we must remember that their throughput is responsible for the water that we depend on for life. Many scientists have worked to establish the water volumes and flows indicated in Figure 3.2, including Raymond Nace (already mentioned in Chapter 2) and the Russian, M.I. L'vovich.

An essential requirement of Figure 3.2 is that of a **water balance** – that, ignoring short-term perturbations, the rate of flow into any of the stores is equal to the rate of flow out of it. Thus the total evaporation from the Earth's surface must, over a long period of time, be equal to the total precipitation. Similarly, the rate at which water arrives on the land areas, as precipitation, must be equal to the total rate at which water leaves them as evaporation, as river flow and as direct groundwater discharge to the seas.

The assumption underlying this is that the total amount of water on

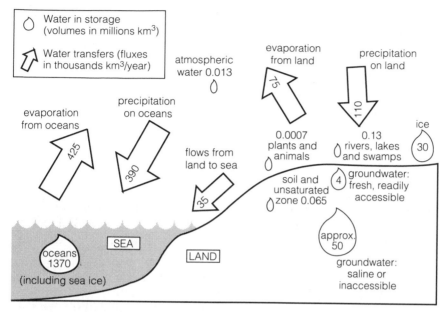

Figure 3.2 A budget for the world's water.

Earth never changes, it merely moves around from one place to another. In general this assumption seems reasonable. Water is a stable substance. Small amounts may be broken down in the upper atmosphere into hydrogen and oxygen, and small amounts are broken down by plants, during photosynthesis, to help form carbohydrates, but the total amount involved in these processes is a negligible fraction of the world's water. The implications of this are that by and large we are drinking and bathing in the same water as the Romans before us, and the dinosaurs before them.

A large amount of precipitation occurs over the oceans, thereby short-circuiting the water cycle. Water that is precipitated over land areas may take several routes through the remainder of the cycle. Some of it will never reach the ground surface. It will be **intercepted** by foliage, and held on the leaves of trees and plants until it evaporates. Coniferous trees are particularly likely to intercept rainfall, partly because of the arrangement of their leaves and partly because those leaves are present throughout the year. Some of the water reaching the ground will fall on bare impermeable rock or on artificial, paved surfaces. Apart from that which collects in depressions and remains there until it evaporates, this water will run off these surfaces into natural or artificial drainage channels. The remainder of the water will fall on soil, and it is largely the condition of that soil which will determine what happens to the water thereafter. Broadly speaking, the water falling on soil may be disposed of in three ways: it may be evaporated, either directly or by transpiration from vegetation after being drawn up by plant roots; it may run over the surface of the soil or travel in the near-surface soil layers until it reaches a ditch or stream; or it may soak into the deeper layers of the soil and so perhaps into the underlying rock.

In the early part of the 20th century, when these processes were under-stood even less well than they are today, it was commonly assumed that the rainfall over a particular area would always be disposed of in more or less the same proportions. It was common to hear engineers talk of the evapo-ration fraction, the runoff fraction, and the percolation fraction. Now that we understand these things more fully – or at least, think we do – we know that the division of precipitation into these end products is more complex.

When rain begins to fall on relatively dry soil, we know from experience that it is readily absorbed. In everyday terms we say that the rain soaks into the ground; in scientific terms we speak of **infiltration**, which is the process whereby water enters the ground at the surface.

If the soil is dry and the rainfall is light, all the water reaching the ground will infiltrate into the soil and be held there as films of moisture which surround the individual soil particles. This water is held in the soil until it is either evaporated directly from the soil surface or taken up by the roots of plants. A small fraction of the water which the roots take up is retained in the plants as part of their growth process, but the majority is evaporated from openings in the leaves and stems in the process known as **transpiration**. The combined effects of evaporation and transpiration in returning water to the atmosphere are frequently grouped together and termed **evapo-transpiration**.

As each successive layer of the soil absorbs water, infiltration moves on downward through the soil and subsoil to the underlying rock. If this rock is permeable, the infiltration process will continue downwards, through the unsaturated zone, until the infiltrating water arrives at the water table and joins the groundwater in the saturated zone. Precipitation reaching the water table is called **recharge**, because it is helping to replenish the store of groundwater.

The maximum rate at which water can enter the soil is called the **infiltration capacity** of the soil. If the rainfall is exceptionally heavy, a situation may arise where water is arriving at the surface of the soil more quickly than it can enter the soil; in this case, the infiltration capacity is said to be exceeded by the rate of rainfall. In these circumstances, the soil will behave like an impermeable surface; depressions will fill, and then **overland flow** (water flowing across the ground surface, usually as small trickles and rivulets) will occur. **Surface runoff** is that part of **total runoff** (the river flow leaving the area) that results from overland flow. This overland flow is sometimes called **Hortonian flow** after R.E. Horton, the American hydro-logist who put forward the theory of infiltration capacity.

In practice, this type of overland flow is rare. In areas with well-developed soils, it occurs only when the soil is frozen or during exceptionally heavy rain. Even in places where vegetation is sparse or where soils are thin or absent, it seems that what begins as overland flow often soaks into soil elsewhere before it can reach a drainage channel as surface runoff. Modern theories actually attribute little or no streamflow to surface runoff, placing much more emphasis on subsurface flow. However, not all the water that has infiltrated the soil reaches the water table. If the rocks beneath the soil are impermeable, or if there are layers in the soil which are themselves of contrasting permeability, there will be a tendency for water to move laterally through the unsaturated zone until it eventually arrives at stream or river channels. This flow is called **interflow**, because it is intermediate between overland flow and true groundwater flow.

Water that has reached the water table becomes groundwater. It **percolates** slowly through the aquifers, at rates which under natural conditions may vary from more than a metre in a day to only a few millimetres in a year. The groundwater moves towards an outlet from the aquifer, which is usually a point where the water table intersects the ground surface. Where this occurs, water will seep or flow from the aquifer; in doing so, it will cease to be groundwater and will revert to being surface water, usually finding its way into a river channel.

The water in rivers usually finds its way back to that primary store of the Earth's water, the sea, but there are some exceptions. Some water will evaporate from the river surface and return to the atmosphere. Under certain conditions, water may flow from the river into the ground, so that rivers may lose water to an aquifer as well as gain from it. There are some parts of the world, such as intermontane basins in semi-arid regions, where the combination of these effects is sufficient to cause rivers to dry up completely, long before they can reach the sea. These are the exceptions, however, to the general Biblical rule that 'All the rivers run into the sea' (Ecclesiastes 1, v. 7), thereby completing the water cycle.

The water cycle is also referred to as the **hydrological cycle**. Hydrology is 'the science that treats of waters of the Earth, their occurrence, circulation, distribution, their chemical and physical properties, and their reaction with their environment, including their relation with living things'. This definition was prepared for the President of the United States in 1962, but hydrology is much older than this implies. We know that people have speculated on the origins of rivers, and on what sustains their flow, since the earliest civilizations. Today, our version of the hydrological cycle seems so logical and obvious that it is difficult to believe that it did not gain widespread acceptance until the 17th century. This was caused in large part by the tendency of the philosophers of Ancient Greece to distrust observations and by the tendency of later philosophers to accept the opinions of the Greeks almost without question. Plato advocated the search for truth by reasoning. He and his followers appear to have attached little importance to observations and measurements. Thus Aristotle, Plato's most famous pupil, was reportedly able to teach that men have more teeth than women, when simple observation would have dispelled this idea. From a hydrological viewpoint, however, he had a more serious misconception – he believed that rainfall alone was inadequate to sustain the flow of rivers.

This error could not be corrected until it was realized that observation and measurement are an essential part of the advancement of scientific knowledge. The first person to make a forthright and unequivocal statement that rivers and springs originate entirely from rainfall appears to have been a Frenchman called Bernard Palissy, who put forward this proposition in 1580. Despite this, in the early 17th century many workers were still in essence following the Greeks in believing that sea water was drawn into vast caverns in the interior of the Earth, and raised up to the level of the mountains by fanciful processes usually involving evaporation and condensation. The water was then released through crevices in the rocks to flow into the rivers and so back to the sea.

In 1674 Pierre Perrault, a French lawyer, published anonymously a book on the origin of springs in which he described how he had compared the annual flow of the headwaters of the River Seine with the amount of precipitation falling each year on its catchment. He found that the precipitation was equal to about six times the flow, and concluded that, in general, 'the waters of rains and snows are sufficient to cause the flow of all the Rivers of the World'.

Edmé Mariotte, another Frenchman and a scientist of great standing, carried out a similar exercise for the whole of the Seine catchment above Paris. He arrived at a similar conclusion, though his work was not published until 1686, two years after his death.

If Perrault and Mariotte demonstrated that precipitation could supply all the flow of the world's rivers, it fell to an Englishman to demonstrate the logic of the other half of the hydrological cycle – the ability of evaporation from the oceans to account for the precipitation. Edmund Halley is best known as an astronomer and as a friend and colleague of Isaac Newton. During astronomical observations from the island of St Helena, Halley was troubled by condensation on the lenses of his telescopes and became so interested that, on his return to London, he carried out a series of experiments. As a result of calculations based on measurements of the rate of evaporation from a pan of water, Halley felt able to conclude in 1693 that the water which evaporates from the Mediterranean Sea on a typical summer day would be three times the amount flowing into the sea on that day in the major rivers. In other words, in general terms, there is sufficient evaporation from the seas to account for river flow.

Perrault, Mariotte and Halley may be described as the first quantitative hydrologists. They established the concept of the hydrological cycle not by speculation, but by observation, measurement and calculation. Groundwater is part of that cycle; since hydrogeology is the study of groundwater, it follows that hydrogeology is a subdivision of hydrology as much as it is of geology. So it is: just as water in its natural cycle takes different forms in different realms – ice, liquid and vapour in glacier, sea, river, aquifer and atmosphere – so hydrology encompasses many fields of study. It is a multi-disciplinary science, involving mathematics, physics, chemistry, meteorology, geology and glaciology, in addition to biology in all its diversity. Hydrogeology must be just as broadly based because, as later chapters will show, it is impossible to isolate completely one portion of the hydrological cycle from the others.

SELECTED REFERENCES

Berner, E.K. and R.A. Berner 1987. *The global water cycle*. Englewood Cliffs, NJ: Prentice-Hall.

Biswas, A.K. 1970. *History of hydrology*. Amsterdam: North-Holland.

Mason, B.J. 1975. *Clouds, rain and rainmaking*, 2nd edn. Cambridge: Cambridge University Press.

Nace, R.L. 1969. World water inventory and control. Chapter 2 of *Water, earth and man* (ed. R.J. Chorley). London: Methuen.

Todd, D.K. 1980. *Groundwater hydrology*, 2nd edn. New York: Wiley.

UNESCO 1971. Scientific framework of world water balance. *Technical papers in hydrology*. Paris: UNESCO.

Vallentine, H.R. 1967. *Water in the service of man*. London: Penguin. Out of print.

Ward, R.C. and M. Robinson 1990. *Principles of hydrology*, 3rd edn. Maidenhead: McGraw-Hill.

4 | Caverns and capillaries

Water is a commonplace and probably, to many people, a rather uninspiring substance. Yet it has a number of unusual properties that not only influence its behaviour underground but also have a great effect on our lives, and that combine to make it unique.

To begin with it occurs on Earth, under natural conditions, in all three physical states – as a solid, as a liquid and as a gas – a fact that is true of no other common substance. When it freezes, instead of contracting like other substances, it expands. This expansion is a nuisance to the householder and to the water engineer, because it can cause water pipes to burst; to the geologist it means that water at about freezing point is a powerful geological agent, because the freezing of water in rock crevices can eventually shatter rocks into fragments.

Another thermal oddity is water's high specific heat – it takes more heat to raise the temperature of a kilogramme of water by one degree than it takes for any other common substance; similarly, water gives up more heat on cooling. This is one reason why the sea has such a stabilizing effect on the climate of adjacent land areas, and why water is so effective in putting out fires.

Water is also one of the most effective solvents known. There are few materials that do not dissolve in it to some extent; each time you drink a glass of water, for example, the water will have a small amount of glass dissolved in it.

Water containing dissolved carbon dioxide is capable of dissolving calcium carbonate, the main constituent of limestone, which is only slightly soluble in pure water (Chapter 11). Water falling as rain dissolves some carbon dioxide on its passage through the atmosphere, and much more when it infiltrates the soil. When rain falls on a region composed of limestone, it will infiltrate and percolate through cracks and crevices, slowly dissolving the rock. If there is no exit for the water, it will become saturated with calcium carbonate and no further dissolution will occur, but if a throughflow becomes established the dissolved material will be carried away and the crevices enlarged by further dissolution. In this way the giant caverns of Somerset, Derbyshire and elsewhere in Britain have been formed; there are more impressive examples elsewhere, like the Mammoth Cave System of Kentucky and the Carlsbad Cavern of New Mexico, which has one chamber more than a kilometre long, 200 m wide and almost 100 m high.

The dissolution of calcium carbonate is a reversible process (the material is actually transported in the form of the bicarbonate, the chemical equation being $H_2O + CO_2 + CaCO_3 \rightleftharpoons Ca(HCO_3)_2$). As the water with its dissolved material moves through crevices and caverns, then evaporation, partial loss of carbon dioxide, or the change in pressure may result in calcium carbonate being precipitated, especially if the cave is only partly filled with water. If precipitation occurs as water drips from the roof of a cave, the result will be a **stalactite** – an 'icicle' of calcium carbonate – growing slowly downwards from the roof. Water dripping from the stalactite may give rise to further carbonate precipitation on the cave floor, so that another structure – a **stalagmite** – will grow upwards. (An infallible way of remembering which way stalactites and stalagmites grow is to think of 'ants in your pants' – 'mites' go up, 'tites' go down!) The two may eventually join, forming a continuous column from floor to ceiling. Clusters of stalactites and stalagmites may take on strange and beautiful forms and are often given fanciful names, especially in caves which are easily accessible to visitors.

Where water runs in sheets down cavern walls, the deposits themselves become curtain-like structures of calcium carbonate in a form called **travertine**, and may look like solid waterfalls. Objects placed in the path of this carbonate-rich water will become covered with calcium carbonate. This phenomenon has led to the advertisement of 'petrifying wells', where one may see everyday objects that have become coated in this way. These have not been petrified in the geological sense, which would mean that each molecule of the original material had been replaced with a molecule of a mineral, but have merely become encrusted.

However, as was said earlier, these limestone caverns and their beautiful structures are a rarity. In most limestone areas groundwater moves through crevices or fissures which, even when enlarged by dissolution, are rarely more than a few millimetres across and which are often less than a millimetre. In other rock types, such as sandstones, it is rare for fissures to be enlarged by dissolution at all, and the permeability and the flow of water are much more diffuse.

In these granular materials another property of water called **surface tension** comes into play. This is not unique to water but is common to all liquids; it arises from the attraction that the molecules of the liquid have for each other, and has the effect of making the liquid surface behave almost as though it were an elastic membrane. In reality of course there is no membrane, but many of the effects that occur at a liquid boundary can conveniently be considered to be the result of a force – surface tension – acting in all directions parallel to the liquid surface. Surface tension has units of force divided by length, so in the SI system it is expressed in newtons per metre (N/m).

One of the effects of this molecular attraction or surface tension is a tendency to reduce the free surface area of any body of liquid to a minimum. The geometrical shape that has the smallest surface area for a given volume is a sphere, so we should expect a mass of water, freely suspended, to assume a spherical shape. We see in drips from taps and in raindrops that, subject to modification by gravity, this is so.

When pure water is spilled on clean glass, the water spreads over the glass in a thin film, tending to wet the glass evenly. This is because the attraction between the molecules of water and the molecules of glass is *greater* than that which molecules of water have for each other; as a result, the water spreads itself over the glass in a layer which, in the limiting case, would be only one molecule thick. Because in this case the surface of the water and the surface of the glass are parallel to each other, we say that the **angle of contact** between the surfaces is zero (Figure 4.1a).

If the sheet of glass is dirty the water does not spread out so evenly; we say that it has less tendency to 'wet' the surface. It no longer tends to form a uniform film, but instead forms discrete globules (Figure 4.1b), each of which has a finite angle of contact θ with the glass surface.

In some cases the molecules of a liquid have a greater attraction for each other than they have for the molecules of the solid with which they are in contact. In these cases the solid appears to repel the liquid; the angle of contact between liquid and solid is greater than 90°, and we say that the liquid is 'non-wetting' to that solid (Figure 4.1c). Examples are mercury on glass, and water on wax. The reason why wax polish helps to protect cars

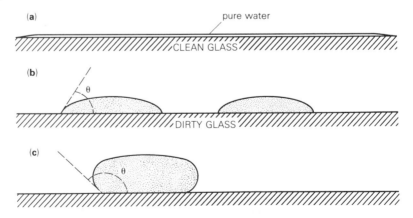

Figure 4.1 Surface tension. (a) On clean glass, pure water spreads out to form a thin film. (b) On dirty glass, water shows less tendency to wet the whole of the surface. It tends to stay in globules, with a finite angle of contact θ between the water surface and the glass. (c) In the case of mercury on glass, or water on wax, the tendency to form globules is very pronounced and the angle of contact θ is greater than 90°. For mercury on glass, θ is about 140°.

is that water is repelled by the wax, and rolls from the surface more easily than it would from the un-waxed paintwork.

A result of surface tension that is of particular importance to us is the phenomenon called **capillarity**. If a glass capillary – a tube with a fine bore – is held more or less vertically with its lower end dipped in water, the water rises up the capillary tube by an amount that depends on the radius r of the tube (Figure 4.2a). The surface of the water comes to rest at a vertical height h above the free surface of the water outside the tube. Also, the water surface inside the capillary is not flat; close examination shows that, like any liquid surface, it is curved where it comes into contact with the solid walls of the tube. We recognize this as a direct result of surface tension; once again the attraction of the glass molecules for the water molecules is greater than that of the water molecules for each other, so that the water is pulled upwards around the circumference of the tube to form the shape we call a **meniscus**. In a wide tube the meniscus will have a flattened central area (Figure 4.2b), but in a narrow, circular tube the meniscus is shaped like a segment of a sphere; if the angle of contact between water and glass is zero, then the radius of this sphere will be the same as the radius r of the capillary tube, and the segment will be a hemisphere.

When the interface between a liquid and a gas is curved, the pressure on the concave side of the interface is greater than that on the convex side. In Figure 4.2a the pressure at A and the pressure at B are equal (they are both

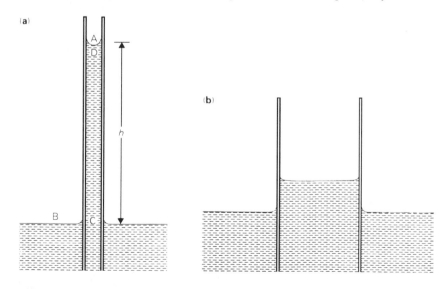

Figure 4.2 Capillarity. (a) In a narrow capillary the air–water interface is hemispherical. The pressure at D is less than the pressure at A and therefore less than the pressure at B, which, like A, is at atmospheric pressure. Water rises up the tube until its hydrostatic pressure compensates for this pressure reduction and the pressures at B and C are the same. (b) In a wide capillary, the interface is not a perfect hemisphere; the central portion is flattened.

atmospheric), and when the water column is at equilibrium the pressure at C must equal the pressure at B. Because A is on the concave side of the meniscus, the pressure at D is less than that at A and B; therefore the height of the water column must be such that the extra pressure which it causes at C is compensated for by the pressure drop across the meniscus. In practice, for pure water in a clean glass tube, this occurs when

$$h = \frac{15}{r}$$

(4.1)

with h and r both measured in millimetres (see Box). A little thought shows that if the water at B and C is at atmospheric pressure, then at all points between C and D its pressure is less than atmospheric.

THE HEIGHT OF THE WATER COLUMN IN A CAPILLARY

In general terms, the pressure drop across a spherical gas/liquid interface is equal to $2\sigma/R$, where R is the radius of curvature of the interface and σ is the surface tension of the liquid. In the case of capillary rise, the pressure deficiency caused by the curvature of the meniscus (Figure 4.2a) must equal the pressure at C caused by the height h of the liquid column; this pressure is $\rho g h$, where ρ is the liquid density and g is gravitational acceleration. Hence

$$\frac{2\sigma}{R} = \rho g h$$

so that

$$h = \frac{2\sigma}{R \rho g}.$$

(4.2)

If the angle of contact is θ, then $R = r/\cos \theta$ so that

$$h = \frac{2\sigma \cos \theta}{r \rho g}.$$

(4.3)

For pure water in clean glass, θ is approximately zero, so that $\cos \theta = 1$ and $R = r$, i.e. the meniscus is a hemisphere with the same radius as the capillary tube. At 10 °C, $\sigma = 0.074$ newton/metre and ρ is approximately 1000 kg/m³. Therefore, taking g as 9.81 m/s²,

$$h = \frac{2 \times 0.074 \times 1}{r \times 10^3 \times 9.81} = \frac{1.5 \times 10^{-5}}{r}$$

with h and r in metres, or

$$h = \frac{15}{r}$$

with h and r in millimetres.

All this may seem a little remote from hydrogeology. Surely rocks do not have long thin cylindrical tubes running through them? They do not, but the interconnected pores in a granular material can behave as though they were bundles of capillary tubes. Let's go back to our substitute sandstone, the lump of sugar. If a lump of sugar is held so that its lower surface is in contact with tea or coffee, the liquid will be seen to be drawn up into the pore spaces in the sugar lump. We cannot observe this for long because the sugar soon begins to dissolve, but the experiment demonstrates that capillary effects do occur in porous materials.

We would therefore expect to find surface tension and capillary phenomena occurring in rocks, and we should not be disappointed. One of the best known is the **capillary fringe** (Figure 4.3). This is a layer of rock immediately above the water table in which water is held by capillarity. Within the capillary fringe the rock is still more or less saturated, but it is distinguishable from the saturated zone proper by the fact that a well will fill with water only to the base of the capillary fringe, i.e. to the water table. Water in the capillary fringe is referred to, not unreasonably, as **capillary water**, to distinguish it from the true groundwater of the saturated zone.

The physical difference between groundwater and capillary water, which explains why wells fill to the level of the water table, can be seen if we refer back to Figure 4.2a. We said that the water between C and D was at a pressure below atmospheric, while that at B and C was at atmospheric pressure. In the ground, water in the capillary fringe (like water in the capillary tube between C and D) is at *less* than atmospheric pressure; below the water table the groundwater is at a pressure *above* atmospheric, in accordance with the laws of hydrostatics. This is illustrated in Figure 4.4. This gives us a way of defining the water table that is more general – and

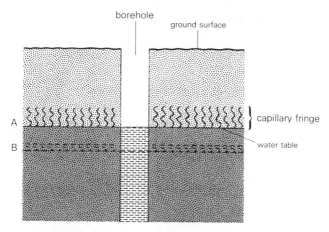

Figure 4.3 The capillary fringe. If the water table is lowered (e.g. by pumping) from A to B, the capillary fringe will also be lowered. If the rock properties at B differ from those at A, then the new capillary fringe may be thicker, or – as in this case – thinner than the original one.

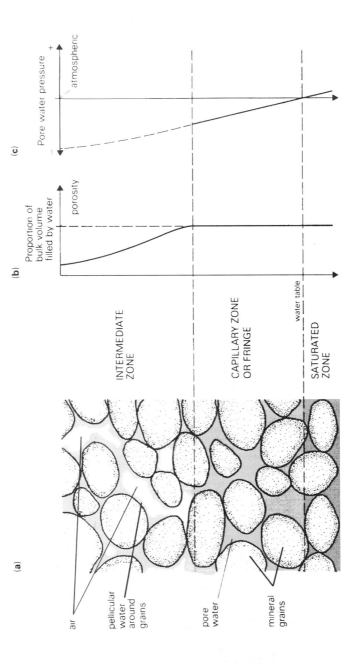

Figure 4.4 Pore water in the unsaturated zone. (a) An exaggerated view of the intermediate zone, with pellicular (film) water around mineral grains, above a capillary zone and water table. (b, c) Graphs of pore-water content (b) and pore-water pressure (c) for the section shown in (a). The precise shapes of the upper parts of the curves in (b) and (c) will depend on previous infiltration and evaporation events. It is important to realize that (a) is a two-dimensional section through a three-dimensional aggregate of grains – this is why in the figure it appears that not all grains are in

therefore more useful – than simple reference to the level at which water stands in a well. We can say that the **water table** is that surface in an underground water body at which the water pressure is exactly equal to atmospheric pressure.

The thickness of the capillary fringe depends on the effective radii of the capillary tubes formed by the interconnected pores of the rock formation – in other words, on the sizes of the connecting 'throats' between adjacent pores. It thus varies from one rock type to another. In a coarse gravel it may be only a few millimetres thick; in chalk or clay it may be many metres. The pores of the material will not be all the same size, so the top of the capillary fringe will not be an abrupt surface but an irregular and gradual transition.

If the water table in an aquifer falls for any reason, the capillary fringe falls with it. However, the rock that was formerly part of the capillary fringe does not drain completely of water; surface tension and molecular effects cause a thin film of water to stay in place around each particle of rock material (Figure 4.4). This is important, because it means that not all the water in the pore space of an aquifer can be abstracted and used. If the water table over an area A of an aquifer falls by an amount z, then the volume of rock affected is $A \times z$. The pore space contained within this volume is $A \times z \times n$, where n is the porosity expressed as a fraction. But the volume of water that will drain from the aquifer is $A \times z \times S_y$, where S_y is less than n and is called the **specific yield** of the aquifer. Specific yield is the ratio of the volume (V_w) of water that will drain by gravity from a rock or soil that was initially saturated to the volume (V_b) of the rock or soil, i.e.

$$S_y = \frac{V_w}{V_b}$$

(4.4)

or

$$S_y = \frac{V_w}{V_b} \times 100\%.$$

(Strictly speaking, all the water that drains from the aquifer in the example above does not come from the volume of rock $A \times z$. Some of the water will have drained from the original capillary fringe above this volume, while some of the water within the volume under consideration will be retained in a 'new' capillary fringe. If the rock properties vary considerably between the two positions of the capillary fringe, this fact may have an important and complicated effect upon the drainage behaviour. See Figure 4.3.)

The water that is unable to drain from the pores is referred to as **specific retention** (S_r), which is the ratio of the volume (V_r) of water that a rock or soil, after being saturated, will retain against the pull of gravity, to the volume (V_b) of the rock or soil, i.e.

$$S_r = \frac{V_r}{V_b}.$$

(4.5)

The definitions of specific yield and specific retention assume that gravity drainage is complete when the ratios are measured.

Because the water that drains and the water that is retained both originally combined to fill the pore space, it should be obvious that the sum of specific yield and specific retention is porosity. In coarse-grained rocks with large pores, the capillary films occupy only a small proportion of the pore space and the specific yield will almost equal the porosity. In fine-grained rocks like chalk, capillary effects are dominant and specific yield will be almost zero.

If you have ever laundered a large article like a blanket or bath towel, you will have experienced one of the effects of surface tension on drainage. You can take a large, wet towel and squeeze out as much water as possible, but within a short time of hanging it on a clothes line the lower edge will be saturated and dripping water. The reason for this is that while the towel is being squeezed or wrung, the vertical capillaries are relatively short, so that their effective vertical length is *less than* that of the column of water which they can support; consequently, they remain full of water. When the fabric is hanging vertically, the capillaries are *longer than* the water column which they can support, so water begins to drain from the bottom of them, and so from the bottom edge of the fabric.

This means that if we take a lump of rock, saturate it with water and leave it to drain, we cannot expect a quantity of water equivalent to the specific yield to drain from it. If the rock is coarse-grained, with correspondingly large pores, *some* water will drain from it, but not the whole of the specific yield. There will always be a layer at the lower side of the sample where, as in the capillary fringe, the pores are completely filled with water; in this layer, in simple terms, the weight of the water is balanced by the retention forces holding the water in the capillary-sized voids. Because of this effect, the determination of specific yield using small samples is difficult, and can require elaborate equipment.

So far, we have seen that we have a saturated zone, whose upper boundary is the water table; above this is the capillary fringe. Both the saturated zone and the capillary fringe are characterized by the fact that all their voids are water-filled.

Above the capillary fringe we have the **intermediate zone** in which the water content is generally at the specific retention value, in the form of thin films surrounding particles and lining the sides of pores; this water is called **pellicular water**. Any excess water in this zone drains under gravity towards the water table.

Above the intermediate zone, which may vary in thickness from zero to more than a hundred metres, is that complex fragment of the Earth's crust, the soil. The capillary fringe, the intermediate zone and the soil are all part of the unsaturated zone, in which surface tension effects play a major role. These effects, and the concepts of specific retention and specific yield, are all relevant to the soil zone, but the relationship between mineral particles and water in the soil is made more complex than elsewhere by the activities of plants. For this reason, and because the soil is the first control on infiltration as it starts its downward movement, the soil deserves a chapter to itself.

SELECTED REFERENCES

Duff, P.McL.D. 1993. *Holmes' Principles of physical geology*, 4th edition. London: Chapman & Hall.

Ford, D.C. and P.W. Williams 1989. *Karst geomorphology and hydrology*. London: Chapman & Hall.

Freeze, R.A. and J.A. Cherry 1979. *Groundwater*. Englewood Cliffs, NJ: Prentice-Hall. (See especially Ch. 2.)

Rodda, J.C., R.A. Downing and F.M. Law 1976. *Systematic hydrology*. London: Newnes-Butterworth. (See especially Ch. 4.)

Tabor, D. 1969. *Gases, liquids and solids*. Penguin Library of Physical Sciences. London: Penguin. (Contains – pp. 212–19 – an excellent presentation of surface-tension and capillarity theory.)

5 | Soil water

To say that the soil deserves a chapter to itself is an understatement. The soil accounts for only the top metre or so of the Earth's crust, but it has probably attracted as much attention as the rest of the Earth's crust put together. Soil physicists and chemists work alongside biologists and agriculturists, and the products of their labours would fill a library, let alone a chapter. The reasons for all this interest are not difficult to find. The soil is our plane of contact with our planet; we live on it, and when we die we become part of it. Plants grow on and in it, and land plants directly or indirectly provide most of our food and oxygen.

The soil is a complicated mixture of organic and inorganic material, a curious blend of life and decay. It consists essentially of mineral particles, dead organic matter, living organisms, water and air. The mineral particles are derived from the decomposition of rocks by the processes known collectively as **weathering**; the rock debris may stay more or less in place or be transported by ice, wind or water. In either case the mantle of rock waste provides a base for living organisms. Bacteria are probably the first to arrive, followed by mosses and lichens. As these die and begin to decay, their remains form the first humus. Seeds are brought by the wind or by animal activity, and a cover of vegetation begins to form. The roots of plants help to bind the mineral particles together, and decaying plant material provides more humus. Earthworms and burrowing animals help to distribute the organic materials through the soil, and provide passageways for water and air. The respiratory activity of micro-organisms increases the carbon-dioxide content of the soil atmosphere, and provides a large proportion of the carbon dioxide which enables rain water to dissolve calcium carbonate.

On steep slopes the rock waste tends to be carried away before plants can establish a foothold to grow and bind it together, which is why such slopes are frequently bare rock or scree. On shallower gradients the type of soil and its rate of development will depend largely on the climate and on the type of rock from which the minerals are derived.

Limestones, as we saw earlier, are dissolved by rain water. Weathering of limestones therefore leaves little in the way of waste; only a few insoluble impurities remain, so that on limestone areas characteristically thin soils are formed. Sandstones weather to sandy soils which are usually permeable and which drain easily. Clays and shales generally weather to give heavy soils of low permeability, which may require careful tilling and extensive artificial drainage to make them productive.

The characteristic feature that turns rock waste into soil is the presence of living organisms, and these need water. Water generally enters the soil by infiltration after rainfall, and occurs as films around the solid particles, held by molecular and surface tension forces (Figure 5.1); air is also present in the soil voids. The thickness of the water films, and therefore the proportion of water to air, depends on the **degree of saturation** of the soil-pore space or on what is sometimes called the prevailing soil-moisture condition. This can vary greatly, depending on the weather.

Suppose that we start with a soil that has just experienced prolonged heavy rainfall, so that all the voids are, unusually, completely filled with water; the soil is saturated. We then allow the soil to drain under gravity (a process that is generally more or less complete in from two to five days, although it is doubtful whether it is ever truly complete). The water that has drained out by the end of this time represents the specific yield of the soil, and the remaining water the specific retention; under this condition the soil is said to be at **field capacity**, which is another way of saying that it is holding all the water which it can hold against gravity.

The drainage of water from the soil goes hand-in-hand with two important effects. First, if we look closely (Figure 5.1b) at the films of moisture around the soil particles in Figure 5.1a, we see that the interface between air and water is curved, with a concave surface presented towards the air; it is in fact a form of meniscus. We saw in Chapter 4 that under these conditions there will be a difference in pressure between the two sides of the interface, with the lower pressure on the convex side. In the soil the air on the concave side is in contact with the atmosphere and must therefore be at

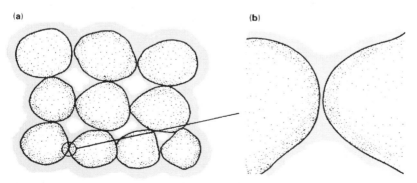

(a) **(b)**

Figure 5.1 Specific retention. Films of water are held around grains by surface tension and by molecular forces.

atmospheric pressure; it follows that the water on the convex side is at a pressure below atmospheric.

This reduced pressure, or suction, holds the water films around the soil grains. It is referred to by various names such as **pore-water suction**, 'soil-water suction' and 'soil-moisture tension'; the last term is especially popular in books on soil physics. As more water drains, the films of water around the grains become thinner and the air–water interfaces become more sharply curved, leading to increased suctions (Figure 5.2). The fact that pore-water suction increases as the thickness of the films reduces means that it is harder for water to drain from smaller pores than from larger ones, so the larger pores drain first. This leads to the second important effect: the larger pores are no longer filled with water and, as it is the larger pores that provide the easier pathways, the soil becomes less permeable to water as it becomes less saturated.

These two effects lead to the effective end of gravity drainage. The reduced permeability of the drying soil to water means that the soil drains ever more slowly; at the same time, the pore-water suctions tend to hold the water around the grains against the pull of gravity. In theory, gravity drainage continues until the gravitational forces are insufficient to overcome the surface-tension forces, or suctions, holding the films of water around the soil particles; in practice, it tends to stop when the films are so thin that significant amounts of water are no longer draining. This is the field capacity condition, the dividing line between specific yield and specific retention – a line that, in practice, is difficult to draw.

In reality, of course, drainage to field capacity does not occur in the ideal way described above, because water is evaporating from the soil surface and plant roots are extracting water from the soil more or less continuously. Once gravity drainage has virtually ceased, the plant roots begin to draw on the specific retention, the moisture content of the soil becomes reduced below the field-capacity figure, and a **soil-moisture deficit** begins to develop. This soil-moisture deficit is usually expressed in terms of the millimetres of rainfall that would have to infiltrate to restore the soil to field capacity.

As the length of time since the last rainfall increases, more and more of the soil water – the specific retention – is used up, either directly as evaporation from the soil surface or by plants. As this happens the films of

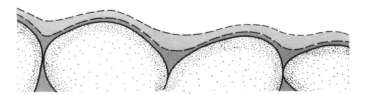

Figure 5.2 Pore-water suction. As water is withdrawn from soil the curvature of the menisci, and therefore the suction holding the water in place, increases as the films of water become thinner.

water around the soil grains become thinner, the larger pores become drained and the air–water interfaces become more sharply curved (Figure 5.2). This means that the pore-water suction increases, and makes it progressively more difficult for plants to continue extracting water, because there is effectively a limit to how hard the plants themselves can 'suck'. To begin with, the plants respond by using less water – the stomata (openings) in their leaves and stems close so that less water is transpired. The reduced throughput of water also means that the plants' rate of growth is reduced.

A balloon maintains its shape because it is filled with air at greater than atmospheric pressure. Herbaceous plants, which have non-woody stems, stay upright because their cells are filled with water at more than atmospheric pressure. If deprived of water for a long period then, just as the balloon goes soft and loses its shape if air is let out, the cells begin to lose their rigidity and the plants begin to **wilt**. Sometimes this happens on very hot days, when the plant is transpiring freely, even though water is still available in the soil; the plant's root system is simply unable to extract water from the soil fast enough to replace that lost by transpiration. In such cases the plant will normally recover overnight, when transpiration stops. Failure to recover in this way means that the wilting is caused not by high demand, but by a restriction in the supply; there is not enough water available in the soil to supply the plant, which is said to have reached **permanent wilting point**. If water becomes available again – perhaps by artificial watering – within a reasonable period of time the plant can recover; if not it will die. Tomato plants display the effects of wilting and recovery very well.

The pore-water suction that exists at the time a plant wilts varies from plant to plant, but typically at the time permanent wilting point is reached the pore-water suction is theoretically capable of supporting a column of water over a hundred metres high. This shows how effective plants are at withdrawing water from the soil.

Under the influence of these high suctions the soil water will redistribute itself. We saw that downward drainage of the soil water ceases early in the drying process; as drying continues and suctions in the root zone increase, water may begin to move upwards to replace the water being removed by evapotranspiration. This upward movement of water will occur first from lower in the soil zone and then from the intermediate zone; in some cases, water may even be drawn up from the saturated zone.

When a commodity is scarce, governments sometimes ration it to ensure that everyone gets a fair share and to encourage people to use no more than they really need. When water becomes scarce, nature seems to impose its own system of rationing. To begin with, downward movement of water to the water table slows as the soil drains to the field-capacity condition. Then, as more of the specific retention is used up, increasing pore-water suctions make it harder for plants to extract water; they are therefore obliged to reduce their consumption by transpiring less and slowing down their growth rate.

However, all this would be to little avail if the high pore-water suctions in the soil zone could simply cause water to move up from the water table; to counter this nature has one more trick up its sleeve. As suctions increase it is the larger pores that are drained first. Eventually water remains in the soil only as thin films held around particles by surface tension or by strong molecular forces. Only the smaller pores within the soil remain completely filled; the larger pores, which provide most of the permeability, are empty and unable to make any effective contribution to water movement. It therefore becomes progressively harder for water redistribution to occur and the upper part of the soil effectively becomes a barrier to further loss of water from the ground.

Having seen something of the way the water in the soil behaves, we can begin to see what characteristics a soil needs to make it ideal for plant growth. Clearly, the soil needs to have a reasonably high porosity, or there will be no space available for water. However, if the soil is coarse-grained, with large pores, those pores will drain easily after rainfall, and there will be little moisture retained for use by plants; hence a high specific yield and low specific retention are undesirable. A high specific retention is needed to retain water against gravity drainage, and this generally means that pores – or the connections between them – must be fairly small. Not too small, though, or the suctions needed to withdraw water from them will be greater than those that plants can exert.

In most soils, significant gravity drainage seems to end when the pore-water suction is around 3 to 5 m. Plant roots can extract water against suctions of more than 100 m, though it gets harder as the suctions increase. Ideally, then, the soil should have pores that will hold water against suctions of 5 m, but release it at suctions of less than 100 m. Not all the pores should be this small, however, because this would mean that the permeability would be low, with the pores remaining filled with water and the surface of the soil becoming waterlogged after heavy rain. This would not do, because although some types of plant can thrive with their roots permanently submerged, most – including those we commonly cultivate for food – need their roots to have access to air as well as water.

Thus the ideal soil for plant growth contains a range of pore sizes, with some large pores, which drain readily and allow air into the soil for plant roots and soil organisms, as well as the many smaller pores that hold water for plant growth. Sands and gravels therefore do not generally provide a good basis for growth unless they are modified in some way. The most effective way of modifying them is to add organic material (humus) that will hold water. Incorporating humus also helps with clay soils by separating the clay particles and providing passages for drainage and air entry.

Most soils contain some pores that are of a much greater size than the pores between the mineral grains of the soil. These larger pores are termed **macropores**, and they often have a biological origin. Common examples are burrows formed by roots that subsequently die and decay, and those formed by burrowing creatures, especially earthworms. At the upper end of the scale are cavities left by tree roots and by burrowing animals such as moles and rabbits.

Pore-water suctions are obviously important to agriculturists and soil scientists. Several methods of directly or indirectly measuring these suctions are available; none of the methods is particularly easy or reliable. In general, the greater the suction to be measured, the greater the problem of measuring it. Many of the methods involve placing a water-saturated block of artificial porous material in contact with the soil. Water will be drawn from the block into the soil until the pore suction in the block is equal to that in the soil; the suction in the block is then measured with some form of vacuum gauge. At high suctions vacuum gauges cannot be used, and indirect methods are resorted to. One popular technique relies on the change in electrical resistance of a porous block as its pore-water suction, and therefore its water content, changes. The block is calibrated in the laboratory by noting the value of the resistance as various known suctions are directly or indirectly applied; measurement of the resistance of the block when placed in the soil enables the suction of the soil to be deduced.

The fact that evapotranspiration is reduced as the soil dries out is important to hydrologists, because evapotranspiration is a major item in the water balance; hence there are various methods for estimating evapotranspiration.

Many methods calculate **potential evapotranspiration**, which is the amount of water that would evaporate, under the prevailing weather conditions, from short vegetation (such as mown grass) completely covering the ground and well supplied with water. These methods make use of measurements such as temperature and wind speed, or of measurements of evaporation from water-filled containers of fixed size and shape. None of the methods is perfect; one will work better under one set of conditions, another under other conditions. Correction factors, which have been determined from field observations, are available to allow us to calculate the evapotranspiration from any other type of vegetation cover if we know the potential evapotranspiration value. Evapotranspiration from a forest of tall pine trees, or from a rice paddy field, for example, will be much more than that from short grass. The hydrologist must consider this when working out the water balance.

On a hot summer day in Britain potential evapotranspiration can be as high as 5 mm/day. On a hot day in North Africa it would typically be about 9 mm/day. (If those figures are less different than you expected, remember that the hours of daylight – and therefore plant growth – are much longer in summer in Britain than in North Africa.)

What happens when the vegetation is not well supplied with water? We have already seen that evapotranspiration falls below the potential value, as the plants begin to cut down on their use of water. A useful concept here is the **root constant** proposed by H.L. Penman; this is the value that the soil-moisture deficit can reach before the plants begin to use less water, i.e. before *actual* evapotranspiration falls below *potential* evapotranspiration. Because some plants are more effective than others at withdrawing water from the ground, the root constant differs from plant type to plant type; it can also vary with the type of soil. For most vegetation, it seems that until the soil-moisture deficit exceeds the root constant by about 25 mm, the rate of evapotranspiration falls only a little below the potential value; thereafter, it declines more rapidly, falling almost to zero as the wilting point is approached.

In practice, plants tend not to follow the rules that we make for them. The agriculturist working out soil-moisture deficits or the hydrologist working out a water balance may assume that a grass area is transpiring at its potential rate when the soil-moisture deficit is, say, 74 mm and at 15% less than its potential rate when the deficit increases to 76 mm; the grass, not being aware of their computer programs, is unlikely to oblige so precisely! We have to realize that concepts like soil-moisture deficit, potential evapotranspiration and root constant are *only* concepts – concepts that we have invented to help us to understand how nature operates, not the rules that nature actually follows.

Many soil scientists now have reservations about the use of concepts like these. Beginning with field capacity, they argue that there is no fixed value of water content or pore-water suction at which gravity drainage of water from pores will completely cease; hence we cannot accurately define specific yield or specific retention. Further, since field capacity is the standard with reference to which we define soil-moisture deficit, it looks as though that must go out of the window too, taking root constant with it. But the test of concepts like these is whether they enable us to make useful and reliable predictions. So long as they help rather than hinder our understanding – which in practice probably means so long as we realize their shortcomings – then these concepts are probably worth retaining, at least until there is something better to put in their place.

Vital as the soil and its water may be to the farmer, what really matters to the hydrogeologist is how much water manages to get through it to recharge the aquifers. Soil is important to us because it is effectively the first control that water meets when it starts to enter the ground. To see this, let us look in more detail at this stage of the water cycle, which was considered briefly in Chapter 3.

Suppose a heavy fall of rain occurs after a long period of dry weather. As rain begins to fall on the dry surface of the soil, the pore-water suctions assist gravity in drawing the water down into the soil, so that it is absorbed readily; the infiltration capacity is very high. As rainfall continues the infiltration capacity will almost certainly fall. The reasons for this include the compacting effect which the raindrops themselves may have on the soil surface, tending to close the pores, and of course the fact that the pore-water suctions will decrease as the pores begin to fill with water. Water will still be absorbed by the soil, however, so long as the infiltration capacity exceeds the rate at which the rain is falling – the **rainfall intensity**.

It may occasionally happen that the rainfall intensity exceeds the infiltration capacity, resulting in the Hortonian overland flow discussed in Chapter 3. It is possible for this to occur even when a high soil-moisture deficit is present – even though the soil needs water, it cannot absorb it fast enough to prevent some water running away over the ground surface. On thick soils, with a cover of vegetation, this process is rarely observed – although it must be said that the heavier the rain, and therefore the more likely overland flow, the less likely there is to be anyone there to observe it! The presence of gullies (water-worn channels) on steep slopes in tropical and sub-tropical areas is evidence that overland flow occurs there; the general absence of such gullies in north-west Europe suggests that it is unusual there. In areas like Britain, the controlling factor on infiltration and the cause of overland flow is likely to occur in the soil profile rather than at the surface: this factor is the rate at which water can move down through successive layers of the soil to leave space for more water to enter the layers above. The effect is most important near stream channels, where the combination of interflow – becoming concentrated as it approaches the channel – and of a shallow water table can result in complete saturation of the soil profile; any further water arriving at the ground surface is thus compelled to travel over the soil as overland flow.

It might seem impossible at first sight that recharge to an aquifer should occur when a soil-moisture deficit exists. It would appear impossible for water to pass downward through soil which has a high pore-water suction until the suctions have been satisfied and the soil restored to field capacity. However, there seems little doubt now that aquifers do receive recharge from rainfall even when soil-moisture deficits are present. One explanation seems to be that many soils crack on drying out. A more general and likely explanation is that infiltration takes place through the macropores described above. Flow taking place through only part of the porosity of a rock or soil is termed in general **preferential flow**, as it is following a preferred route. In the soil and unsaturated zone the macropores, whether natural cracks or pores of organic origin, provide a path for rain water to enter the deeper layers of the soil or intermediate zone, bypassing the shallower layers where a soil-moisture deficit may exist. Preferential flow in the unsaturated zone is therefore often referred to as **bypass flow** because the water is bypassing material that would otherwise retain or delay it.

Rainfall that is not heavy but gentle and prolonged provides ideal conditions for recharge to occur. The rainfall intensity is low enough not to exceed the infiltration capacity nor to cause interflow; then, provided that the underlying materials are permeable, once the soil-moisture deficit is satisfied (or even sooner if cracks or macropores are present), water can move down to recharge the aquifer. What it finds when it gets there – well, that's the subject of the next chapter.

SELECTED REFERENCES

Childs, E.C. 1969. *An introduction to the physical basis of soil-water phenomena.* London: Wiley.

Doorenbos, J. and W.O. Pruitt 1992. *Crop water requirements*, Irrigation and Drainage Paper No. 24. United Nations Food and Agricultural Organization.

Duff, P.McL.D. 1993. *Holmes' Principles of physical geology*, 4th edition. London: Chapman & Hall.

Freeze, R.A. and J.A. Cherry 1979. *Groundwater.* Englewood Cliffs, NJ: Prentice-Hall. (See especially pp. 39–44 and Ch. 6.)

Marshall, T.J. and J.W. Holmes 1988. *Soil physics*, 2nd edn. Cambridge: Cambridge University Press.

Rodda, J.C., R.A. Downing and F.M. Law 1976. *Systematic hydrology.* London: Newnes-Butterworth. (See especially Ch. 4.)

Rowell, D.L. 1993. *Soil science – methods and applications.* Harlow: Longman.

Ward, R.C. and M. Robinson 1990. *Principles of hydrology*, 3rd edn. Maidenhead: McGraw-Hill.

Groundwater in motion | 6

So far in this account I have talked about water entering aquifers, moving through them and leaving them, without really discussing how this process occurs. Now it is time to consider groundwater movement, and its causes, in more detail.

Although this chapter is far from a rigorous treatment of groundwater flow, it does go into more detail than most of the others in this book. I make no apology for this, as it seems to me that an understanding of these basic physical principles (even in elementary terms) is essential to an understanding of hydrogeology. I have summarized the most important points in simplified form at the end; if you find the chapter heavy going, read the Summary. This should give you enough information to understand subsequent chapters. You may then want to return to the more detailed sections afterwards.

WATER MOVEMENT

To understand how groundwater moves we first need to think about the movement of water in general – indeed, we need to think about what makes anything move at all. If you feel like getting up and running about, you might say that you are feeling energetic; if on the other hand you felt particularly listless you could say that you were lacking energy. Energy is simply the capacity to do work. A crude way of measuring how energetic you feel is to see how many flights of stairs you can climb – in other words, how high you can raise yourself above a given level. Climbing stairs is a form of work.

Water, like people, needs energy to make it move; one way of expressing the energy of water, like that of people, is to measure the height to which it can raise itself above an arbitrary given level or **datum level**. We call this height the **hydraulic head** or simply the **head** above that datum. The datum can be arbitrary because, as we shall see, we are mainly interested in comparing the energy of water in one location with that in another.

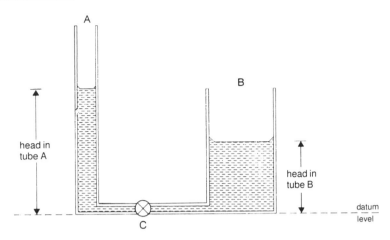

Figure 6.1 Head differences. There is a head difference between the water in tubes A and B. The *head* of water in A is greater than that in B, even though B contains a greater *volume* of water. If valve C is opened water will flow from A to B.

Provided that we measure the energies relative to the *same* datum, the position of the datum itself is irrelevant.

Consider the arrangement in Figure 6.1. Tube B contains more water than tube A, but experience tells us that if we open valve C water will flow from A into B, until the water level in both tubes is at the same height above the datum level. We know that flow occurs from A to B because initially the water level in A is higher than in B. In everyday terms we speak of 'water finding its own level'. More scientifically we can say that there is a **head difference** between A and B, and that the head at A is greater than that at B. Flow occurs to equalize these heads. Is head, then, simply related to height above the ground? In other words, does water always flow downhill? Unfortunately, no – or perhaps I should say fortunately, no, because if water could not flow uphill we should have great difficulty in distributing it around the countryside to meet domestic, industrial and agricultural demand.

Consider the simple case of the water supply to a house (Figure 6.2). The water supply pipe is underground, yet the water travels up the rising main to the taps or into the storage tank, which (in Britain) is usually in the roof cavity. Clearly this could not occur if head were simply a function of height. The upward flow takes place, of course, because the water in the pipe is under pressure. If we insert a **manometer** (an open-topped pipe, of large enough diameter for capillary effects to be negligible) into the water main at point A, water will rise up the pipe to point B to give us a measurement of the **pressure head** at A. So long as B is above the level of the ball valve, water can flow up into the storage tank, and to any taps or appliances.

We therefore see from Figure 6.2 that a head difference can be a result of *pressure* difference, as well as a result of a difference in elevation. To

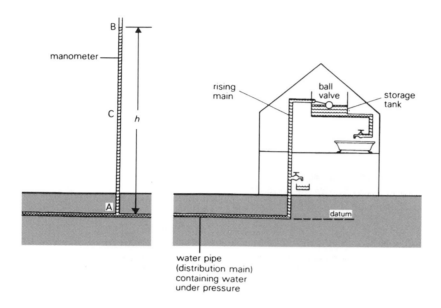

Figure 6.2 Water supply to a house. Water will flow from the underground supply pipe up the rising main to the tap and storage tank because it has pressure energy. The height *h* to which it would rise in a manometer inserted at A is a measure of its 'pressure head'.

distinguish between them, it is quite in order for us to speak of 'elevation head' and 'pressure head'. But are these two types of head really different? To answer this question, we need to go back a little, to the idea of head as a measure of energy.

We can define hydraulic head of water as the energy of the water per unit weight. Water can possess energy in several ways; one of these ways is **elevation energy**, which is energy possessed by virtue of position (Figure 6.3). A mass *m* a vertical distance *h* above a datum has elevation energy *mgh* units relative to that datum – *mgh* units is the amount of work needed to lift the mass from the datum to its position, or the amount of work which we could make it do in allowing it to move from that position to the datum. (As usual, *g* is the acceleration due to gravity.) A unit volume of water has a mass of ρ, where ρ is the density of water; a unit volume of water at height *h* therefore has elevation energy ρgh units. Since the weight of a unit volume of water is ρg, it follows from our definition of hydraulic head (energy per unit weight) that water at height *h* above the datum is at an elevation head of $\rho gh/\rho g$, or *h*. All of which may seem a complicated way of stating the obvious, but it is sometimes advisable to make sure that what is 'obvious', is really based on fact. (It used to be 'obvious', for example, that the Sun moves around the Earth). The other thing that our calculation has revealed is that head has units of length: we can measure it, for example, in metres.

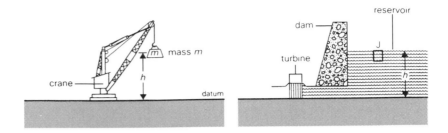

Figure 6.3 Elevation energy. A mass *m* at a height *h* above a datum possesses elevation energy of *mgh* units relative to that datum. This is the work that a crane would have to do to lift the mass from the datum (in this case, ground level) to that height: it is also the amount of work that the weight could be made to do in returning from height *h* to the datum. The water in a reservoir also possesses energy. The amount of work that each unit volume of water can be made to do in passing through the turbine is proportional to *h*. A unit volume of water has mass ρ, so the unit volume at J has elevation energy of ρgh. Its weight is ρg, so its *head* (energy/unit weight) is simply *h*.

Let us now think about pressure. We saw from Figure 6.2 that when we insert a manometer tube into our water main at point A, the water rises up the tube. Clearly then, water under pressure possesses energy and therefore hydraulic head. What we have to do is express this in quantitative terms.

In Figure 6.2, when we have connected our manometer and when the water is no longer flowing, we have a situation in which the water that has moved from A to B has exchanged the pressure energy it possessed at A for the elevation energy it possesses at B. We know from the law of the conservation of energy that these two quantities of energy must be equal; since the water at B has energy ρgh per unit volume and therefore has a head of *h* units, then the water at A must have the same energy and therefore the same head, *h*. If this were not so, flow would take place in the manometer.

In general, the pressure at any depth *z* below the surface of a liquid is higher by an amount *p* than the atmospheric pressure acting on the liquid surface, where

$$p = \rho gz.$$

For convenience, we can adopt a convention of measuring all pressures relative to atmospheric pressure. Thus in Figure 6.2 the pressure *p* at A is equal to ρgh because A is at depth *h* below the water surface. Since the head at A equals the head at B, which in turn equals *h*, we see that the head at A is given by

$$h = \frac{p}{\rho g}.$$

This is a general formula for the calculation of pressure head.

Pressure energy and elevation energy are both forms of **potential energy**, which in general terms is energy that a body possesses by virtue of its position or state. Water can possess both these forms of energy simultaneously; the water at point C in the manometer tube in Figure 6.2, for example, possesses some elevation energy as a result of its height above the datum and some pressure energy as a result of its depth below the water surface.

Elevation energy is potential energy possessed by virtue of position. Pressure energy is potential energy possessed by virtue of state; it is somewhat analogous to the energy of a compressed spring. A clockwork toy, set to run slowly up a slope, converts the potential energy of its spring to elevation energy. Similarly, in Figure 6.2, water flowing from the water main to the storage tank converts its pressure energy into elevation energy.

In any body of standing water, the pressure energy increases with depth below the water surface and the elevation energy increases with height; the sum of the two is constant – if it were not, the water would be in motion. These relationships for static water in a reservoir are shown in Figure 6.4. At any point K, the sum of pressure head h_p and elevation head h_e is equal to the static head h_s, which is the height of the water surface above the datum.

There is a third way in which water can possess energy – by virtue of movement. Moving water clearly possesses energy – a horizontal jet of water, for example, can do work by rotating a turbine wheel. This energy

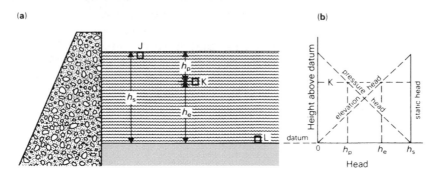

Figure 6.4 Pressure head and elevation head in standing water, such as a reservoir. (a) A unit volume of water at the water surface (e.g. volume J) possesses elevation head but no pressure head; at the datum level (which in this example is at the bottom of the reservoir) a unit volume of water possesses pressure head but no elevation head (e.g. volume L). At any intermediate depth the water (e.g. K) possesses some pressure head h_p and some elevation head h_e. The sum of the two components is the static head, h_s, and is equal to the height of the water surface above the datum. (b) A graphical representation of the head conditions in (a). The elevation head decreases with depth, whereas the pressure head increases with depth; at any depth the sum of the two is equal to the sum at any other depth and to the static head, h_s. Because pressure increases with depth the dam is made thicker towards its base.

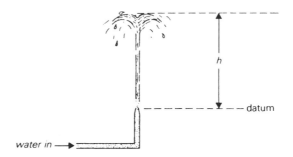

Figure 6.5 Energy changes in a vertical jet of water. On leaving the nozzle, water possesses no elevation energy relative to the datum, but does possess some kinetic energy. At the top of the jet the water's velocity (and therefore its kinetic energy) is momentarily zero but its kinetic energy has been converted into elevation energy. Pressure energy (the water is at atmospheric pressure from the time it leaves the nozzle) is unchanged throughout.

is called **kinetic energy** and the contribution which it makes to the total hydraulic head is called the **velocity head** or **dynamic head**.

A mass m moving with speed v has kinetic energy of $\frac{1}{2}mv^2$. From our definition of hydraulic head, the velocity-head contribution of unit volume of water (mass ρ) is therefore $\frac{1}{2}\rho v^2/\rho g$ or $v^2/2g$. We can demonstrate this using, once again, the law of the conservation of energy.

Suppose we direct a jet of water vertically upwards (Figure 6.5). The water leaves the jet with speed v, and we draw our energy datum line level with the nozzle. Once the water leaves the nozzle it is at atmospheric pressure, so there are no pressure-energy changes. Where it leaves the nozzle it is level with the elevation-energy datum, so it possesses only kinetic energy, $\frac{1}{2}\rho v^2$ per unit volume. At the top of the jet, the water is effectively stationary, so its only energy is elevation energy, ρgh per unit volume. Assuming that no energy is used in overcoming friction, for example with the air, then these two quantities must be equal, i.e.:

$$\frac{1}{2}\rho v^2 = \rho gh$$

whence

$$h = \frac{v^2}{2g},$$

which is what we wanted to demonstrate.

HEAD LOSS

Suppose now that we set up a horizontal long straight length of pipe, whose cross-section is the same throughout its length (Figure 6.6). Near each end

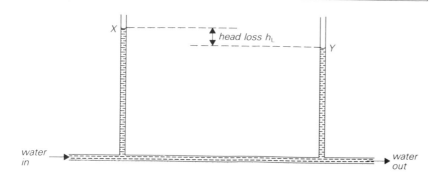

Figure 6.6 Steady flow through a horizontal pipe. The head loss between X and Y represents energy that has been converted into heat.

we fit a manometer. We then pass a steady flow of water through the pipe – by steady, we mean that in any instant the same volume of water enters the pipe as in any other instant, and the volume leaving the pipe in any instant is the same as the volume entering it.

If we carefully observe the water levels in the manometers, we shall see that the water level in the manometer nearest the outflow end of the pipe is lower than the level in the manometer at the inflow end. Surely, at first sight at least, this is a paradox? The pipe is horizontal, so there can be no change in elevation energy as the water flows along it; the flow is steady and the cross-section is uniform, so the speed and therefore the kinetic energy must also be constant. Yet there is a reduction in head, which can therefore only be the result of a decrease in pressure. The decrease in pressure energy has not been balanced by an increase in kinetic or elevation energy, so where has this energy gone? We know from the law of energy conservation that it cannot have been destroyed.

The answer is that the energy has been converted into heat, as a result of friction between the water molecules themselves and – to a lesser extent – between the water and the surface of the pipe. (In basic terms heat is really a form of kinetic energy – when a substance is heated its molecules vibrate more rapidly. It is therefore easy to see how friction – which can be thought of as an interaction between molecules – increases molecular kinetic energy and therefore generates heat.) In just the same way that you become hot if you run around for a time, or that a bearing in a motor becomes warm with use, so the water and pipe will have become heated, although usually by so small an amount that we should not be able to measure the resulting rise in temperature. The sad thing about the energy that has been converted into heat in this way is that it is irrecoverable. We cannot, for example, cool the pipe down and increase the pressure of the water again. Therefore although strictly speaking no energy has been destroyed, *useful* energy has been 'lost' from the system, and for this reason the fall in head between the two manometers in Figure 6.6 is commonly referred to as **head loss**.

The faster you run, the more heat you generate within your muscles; the faster we move water through a pipe, the more energy is dissipated as heat and the greater the head loss along the pipe. Head has units of length, and we can divide the head loss between the ends of the pipe by the length of the pipe to obtain a quantity called the **hydraulic gradient**, or **head gradient**. Other things being equal, the faster we want the water to move, the steeper this gradient must be. In one way water always does flow downhill, but the 'hill' it flows down is the hydraulic gradient – the steeper this gradient, the faster the water will flow. If, in Figure 6.6, we stop the flow altogether the gradient becomes zero – in other words the heads in the manometers become equal; the water 'has found its own level' again.

If we were to close off the outflow end of the pipe in Figure 6.6 the water levels in the manometers would rise; whether or not they actually over-flowed would depend on whether or not they were higher than the head at the inflow point. Let us assume for a moment that the manometers *are* higher, so that now we have a condition of no flow through the pipe, with the heads in the manometers equal. Why are these heads now higher than they were when water was flowing through the pipe? To answer this, we look again at the energy conditions.

When water was flowing through the pipe, in addition to some energy being dissipated as head loss, the flowing water possessed kinetic energy. The manometers do not measure total head, but **static head**, which in turn is proportional to potential energy. When the water begins to move, some potential energy is converted into kinetic energy; the *total* head (except for that part dissipated as heat) remains unchanged, but the *static* head decreases.

We see this very clearly if water flows steadily through a horizontal pipe containing a constriction (Figure 6.7). For water to flow through this constriction at the same rate as through the rest of the pipe, it has to speed up, so that the volume flowing through the constriction in unit time is the same as the volume flowing through any other part of the pipe in unit time. This increase in speed means an increase in kinetic energy, which in turn must mean a reduction in potential energy. We measure this as a reduction in static head between manometer A and manometer B, but in this case only part of the head loss is due to friction, the remainder being caused by the change from potential to kinetic energy.

The elevation energy of the water is constant, so the reduction in potential energy must result from a decrease in pressure. It may seem remarkable that pressure should decrease inside a constriction, but it is a fact. The clockwork-toy analogy may be useful again here. If we set our toy running quickly along a horizontal surface, the spring runs down as its energy is converted into kinetic energy and used to overcome friction. If we wished the toy to go faster, the spring would have to wind down more rapidly to supply the extra kinetic energy; the energy in the spring is analogous to the pressure in the pipe. This pressure change in water flowing

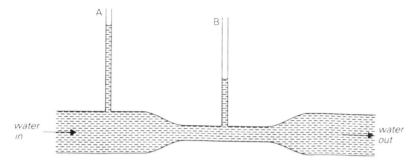

Figure 6.7 Pressure decrease at a constriction. When water flows through a constriction its velocity (and therefore its kinetic energy) increases. If, as in this case, the pipe is horizontal, then the elevation energy is unchanged; therefore pressure must decrease to compensate for the increase in kinetic energy. This pressure reduction is seen as a head difference between manometers A and B.

through a constriction provides a means of measuring the rate of flow of water, using an instrument called a Venturi meter.

On first reading, all this may seem rather complex. The situation looks worse when we stop to think that so far we have been talking mainly about water flowing through a smooth, circular pipe; how much worse things must be when we start to deal with water flowing through the complex twists and turns that make up the pore space in, say, a sandstone! Paradoxically, things are both worse and better. If we tried to describe the water movement along all the individual flow paths between the sand grains, then we should find the task beyond us. Fortunately, we have no need to do this; what interests us is the total flow resulting from water movement through the combination of all these flow paths: this is much simpler. Another simplifying fact is that groundwater generally moves very slowly – so slowly that its kinetic energy (and therefore its velocity head) is usually negligible.

HEAD AND POTENTIAL

More advanced books on hydrogeology use the term **fluid potential**. This is the energy of the fluid per unit mass, and it is given the Greek letter φ (phi) as its symbol. Since head is energy per unit weight, the relationship between the two is simply

$$\varphi = gh.$$

Fluid potential, like head, has a component derived from elevation (gz), a component derived from pressure (p/ρ) and a component derived from motion ($v^2/2$). Thus the total fluid potential is

$$\varphi = gz + \frac{p}{\rho} + \frac{v^2}{2}.$$

Water can therefore be regarded as moving down a potential gradient in just the same way as down a head gradient. The potentiometric surface (Chapter 2) takes its name from the fact that it represents the potential of the groundwater in the aquifer; equally, of course, since potential is proportional to head, the potentiometric surface represents the head.

DARCY'S LAW

In studying the flow of water or of any other fluid through rock, we make use of a law, called Darcy's law, formulated more than a hundred years ago by a Frenchman. Henri Darcy was born in Dijon in 1803 and trained as an engineer. Dijon had long suffered from the lack of a dependable supply of safe drinking water, and shortly after graduating Darcy began work on an attempt to solve the problem. He designed a collection system for water from a large spring more than 10 km from Dijon, piped the water to the city and arranged for its distribution to standpipes. For the first time in its history, Dijon had a reliable water supply.

This practical success alone was a considerable achievement, but Darcy went further. In Paris he carried out a great deal of scientific work on the problems of water distribution, including studies of pipeflow and the measurement of rate of flow. An account of his work at Dijon, together with many of his experimental findings, was published in 1856 under the title *Les fontaines publiques de la ville de Dijon* – the public water supply of Dijon. One chapter dealt with clarifying water by filtering it through beds of sand. It was during the course of this work that Darcy derived the relationship that bears his name.

If we wished to test the relationship for ourselves, we could use an arrangement like that shown in Figure 6.8. Here we have a cylinder of cross-sectional area A, filled with sand which is held between two gauze screens. Water flows through the cylinder at a steady rate Q, measured for example in litres per day. Manometers at X and Y, a distance l apart along the cylinder, measure the head of the water relative to the datum level; the difference between the two heads is the head loss, h_L.

If we double the flow rate Q we find that h_L doubles also – we have to use twice as much energy to drive the water through the sand twice as fast. If we halve Q, we halve h_L; in general we find that the flow rate is directly proportional to the head loss, i.e.

$$Q \propto h_L.$$

Suppose we now vary the length of our sand column, but keep the same head loss between the ends. We find that as l increases, Q decreases, and vice versa, in strict proportion. If we double l, we halve Q. In general,

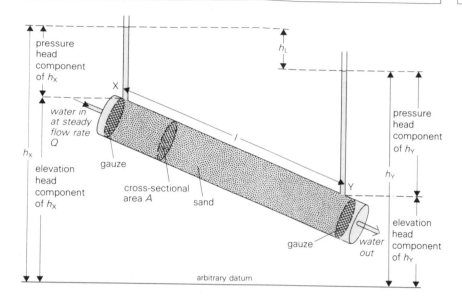

Figure 6.8 Experimental verification of Darcy's law. The flow rate Q is proportional to the cross-sectional area A and the head loss h_L, and inversely proportional to the length l, i.e. it is proportional to A and to the hydraulic gradient h_L/l. The head loss h_L is the difference in head between the two manometers X and Y, which are at distance l apart (l is measured along the direction of flow). The head h_x at X consists of an elevation-head component and a pressure-head component; the same applies to h_y at Y. Notice that in this example the water is flowing from a region of lower pressure to one of higher pressure.

$$Q \propto \frac{1}{l}.$$

Combining these two results, we see that

$$Q \propto \frac{h_L}{l}.$$

We met the term h_L/l when discussing Figure 6.6 and called it the hydraulic gradient. We see therefore that the rate at which water flows through sand is proportional to the hydraulic gradient.

If we put another identical cylinder beside the first, filled with identical sand and with flow occurring in the same way, we should have in effect twice the flow occurring through twice the cross-sectional area. It seems reasonable to assume, as Darcy did, without actually performing the experiment that

$$Q \propto A.$$

In summary therefore we have

$$Q \propto A \frac{h_L}{l}$$

which is the same as saying that

$$Q = KA\frac{h_L}{l}$$

(6.1)

where K is a constant of proportionality.

Since we can measure Q, A, h_L and l in our experiment in Figure 6.8, we can calculate the value of K. If we did this, and then emptied all the sand out of the apparatus, replaced it with finer sand, repeated the experiment and then calculated K again, we should almost certainly find (as Darcy did) that the new value of K was smaller than the first one. If we used coarser sand, we should find that K was larger. K is the permeability, or more strictly the **hydraulic conductivity**, of the sand; coarse sand is more permeable than fine sand because the spaces between coarse grains are correspondingly larger and allow an easier passage for the water.

Equation 6.1 is **Darcy's law**, which, expressed in simple terms, says that a fluid will flow through a porous medium at a rate which is proportional to the product of the cross-sectional area through which flow can occur, the hydraulic gradient and the hydraulic conductivity.

If we repeated our experiments with the sand-filled cylinder but used syrup, say, we should expect, other things being equal, to find the flow rate much less than when we used water. This would mean, since everything else was unchanged, that the value of K that we calculated using Darcy's law

---

DARCY'S LAW AND OTHER LAWS OF PHYSICS

You may recognize the similarity between Darcy's law, Fourier's law of heat transfer, and Ohm's law. Ohm's law is usually expressed as

$$I = \frac{V}{R}$$

where I is electric current, R is resistance and V is the potential difference (the 'voltage') across the resistor. If we consider a cylindrical wire, length l, cross-sectional area A and conductivity c, then

$$R = \frac{l}{cA}$$

so that

$$I = cA\frac{V}{l},$$

and the similarity with Darcy's law is obvious. Clearly current is analogous to rate of flow of water, and potential difference V is analogous to head loss h_L. This analogy is used in resistance network models of aquifers, which will be discussed in Chapter 10.

would be lower. Clearly, then, the hydraulic conductivity K depends not only on the porous medium but on the fluid that is filling the pores and passing through them – a fact that Darcy probably understood, as he described K as 'dependent on the permeability of the [sand] bed'.

The property of the fluid that affects hydraulic conductivity is the kinematic viscosity, which is usually denoted by the Greek letter v (pronounced 'new'); as the kinematic viscosity of the fluid in a porous medium increases, the hydraulic conductivity decreases. The kinematic viscosity of water decreases as its temperature increases.

Most rocks are not uniform, or **homogeneous**. As a result of natural variations during their formation and subsequent alteration, they are **heterogeneous**: their properties vary from place to place within the rock. Figure 6.9a shows an example of vertical heterogeneity. At any point in the rock, the permeability will usually also vary with direction; in Figure 6.9b, for example, it is easier for water to flow between the grains in the *horizontal* direction (parallel to the bedding) than in the vertical direction. Equality of properties in all directions at a point is **isotropy**; variation of properties with direction is **anisotropy**. Most sedimentary rocks are anisotropic with respect to permeability because they contain grains that are not spherical but are elongated in one direction or shortened in another. When deposited, these grains tend to settle with their shortest axes more or less vertical, and this can cause the permeability parallel to the bedding to be greater than that perpendicular to the bedding, by a factor that typically varies from one to four. When heterogeneity like that shown in Figure 6.9a is also taken into account, the bulk permeability (i.e. the permeability of a large block) of sediments is usually much greater parallel to the bedding than perpendicular to it.

In order to talk about relative values of hydraulic conductivity, we clearly need to define it and give it some units. If we consider a homogeneous and isotropic porous medium with all the pores filled with a single fluid, the

(a) **(b)**

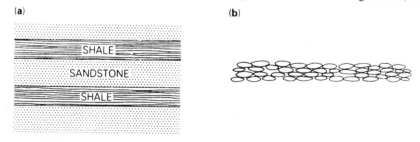

Figure 6.9 Heterogeneity and anisotropy. (a) An example of vertical heterogeneity. It is easier for water to flow between the shale layers than across them. Therefore, although the sandstone and shale layers may individually be isotropic, the horizontal permeability of the combination of beds is greater than the vertical permeability. (b) Within a single rock type, platy grains tend to be deposited with their longer axes horizontal. This texture causes permeability anisotropy although the rock may be homogeneous.

HYDRAULIC CONDUCTIVITY AND INTRINSIC PERMEABILITY

The concept of hydraulic conductivity is a useful one in most topics of hydrogeology, since the viscosity of groundwater (which depends mainly on its temperature and the material dissolved in it) does not vary greatly from one aquifer to another. When considering water at great depths – where it is usually hot and saline – or rocks that contain fluids other than water, it is useful to be able to separate the properties of the rock from the properties of the fluid. This is particularly true when considering gas- or oil-bearing strata, since the viscosities of these fluids are very variable. For this reason the concept of **intrinsic permeability** was introduced (symbol k). The relation between hydraulic conductivity (or field permeability) and intrinsic permeability is

$$k = \frac{Kv}{g} = \frac{K\mu}{\rho g}$$

where v is fluid kinematic viscosity, μ is dynamic viscosity, ρ is density and g is gravitational acceleration. k thus has the dimensions of L^2 or area. A variety of units such as the cm^2 or μm^2 is possible, but the oil industry uses a unit called the **darcy** (plural *darcys*), defined as the permeability that permits a flow of 1 ml of fluid of 1 centipoise viscosity completely filling the pores of the medium to flow in 1 s through a cross-sectional area of 1 cm^2 under a gradient of 1 atm/cm of flow path.

Despite the rather appalling combination of inconsistent units, and the use of a pressure-gradient term instead of the hydraulic gradient that should strictly have been employed, the darcy and its derivative, the millidarcy (1 darcy = 1000 millidarcys), have become the almost universal units of permeability in the oil industry.

When we use the term 'permeability' we are, strictly, talking of k, not K. But in practice, in hydrogeology, the variation in viscosity is usually so small as to make the distinction relatively unimportant. When we speak of one rock as being more permeable than another, it is tacitly assumed that they contain groundwater with similar properties and not that one is filled with water and the other, say, with syrup.

hydraulic conductivity is the volume of the fluid at the prevailing kinematic viscosity that will move in unit time under a unit hydraulic gradient through a unit cross-sectional area perpendicular to the flow. If we use Figure 6.8 as an example, unit hydraulic gradient means that h_L and l must be equal; then, if we make A equal to 1 m^2, the volume rate of flow through the cylinder is equal to the hydraulic conductivity (from Equation 6.1). Hydraulic conductivity thus has units of volume per time per area; hydrogeologists usually express it in terms of cubic metres per day per square metre, which reduces to metres per day (m/day). Civil engineers frequently use the centimetre/second (cm/s); the metre/second (m/s) is sometimes used, but is too large for most purposes.

The permeabilities of natural materials vary by many orders of magnitude. (An order of magnitude is a factor of ten.) Rocks that have hydraulic conductivities of 1 m/day or more when filled with ordinary groundwater are generally regarded as permeable and likely to form good aquifers; those with hydraulic conductivities of less than 10^{-3} m/day would generally be regarded as impermeable. However, all things are relative. A layer of sand with a hydraulic conductivity of 0.1 m/day might be regarded as impermeable if it separated two gravel strata with hydraulic conductivities of 50 m/day, but a similar sand occurring within clay with a hydraulic conductivity of 10^{-4} m/day might be regarded as permeable. A hydraulic conductivity of 10^{-3} m/day might be embarrassingly high to a civil engineer if it occurred in bedrock underneath a dam, because the high head of water behind the dam might cause significant flow through the rock.

It is worth emphasizing at this point that in using Darcy's law we are, as was said earlier, forgetting about the tortuosity of the individual pore channels. We are instead assuming that we can carry out a form of averaging, replacing the twisting microscopic channels of the **porous medium** formed by our rock fabric with a macroscopic **continuum**, through which the flow will be the same as the resultant of all the flows through the microscopic channels of the rock. This may seem, intuitively, a logical thing to do – after all, we are in general interested in the bulk movement of water through the rock, not the minute flow through an individual pore channel – but to demonstrate the validity of the assumption in rigorous physical terms is far from straightforward. We can demonstrate Darcy's law in the laboratory, but to derive it mathematically from the fundamental equations of fluid mechanics is another matter. M. King Hubbert, an American geologist who did a great deal of pioneering work on the movement of groundwater and the migration of petroleum, appears to have been the first person to attempt the derivation with any degree of success – in 1956, to mark the centenary of Darcy's publication. Despite the attempts of Hubbert and others, Darcy's law remains essentially an empirical law, its validity resting on experimental evidence.

APPLICATIONS OF DARCY'S LAW

How is Darcy's law used? In many ways. One of the most frequent and easiest to understand is in calculating the natural flow through an aquifer. Imagine a strip of aquifer (Figure 6.10a), with flow occurring in the direction from borehole X to borehole Y. The hydraulic conductivity is K m/day. How much water flows through in a day?

The cross-sectional area A is simply bw. If the flow is so slow that the kinetic energy of the water is negligible, the hydraulic gradient can be determined from the difference in water levels in the two boreholes, which are effectively manometers. This difference is H, so the hydraulic gradient is H/l, and we see that this is simply the slope of the potentiometric surface.

From Darcy's law, the flow Q is given by

$$Q = KAH/l = KbwH/l.$$

Obviously, the effectiveness of a rock stratum as an aquifer depends not only on its hydraulic conductivity but on its cross-sectional area perpendicular to the flow direction – i.e. on its thickness b and width w. In particular, in carrying water to a well it is the product of hydraulic conductivity K and thickness b that is important. This product Kb, which is frequently given the symbol T, is called **transmissivity**. For the case shown in Figure 6.10a, Darcy's law can therefore equally well be written as

$$Q = TwH/l.$$

Transmissivity has units of hydraulic conductivity multiplied by thickness;

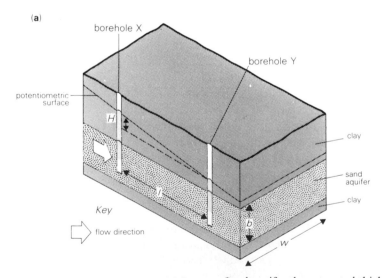

Figure 6.10 Flow through aquifers. (a) In a *confined* aquifer the saturated thickness b is constant. The flow rate Q through the aquifer is (from Darcy's law) equal to $KbwH/I$.

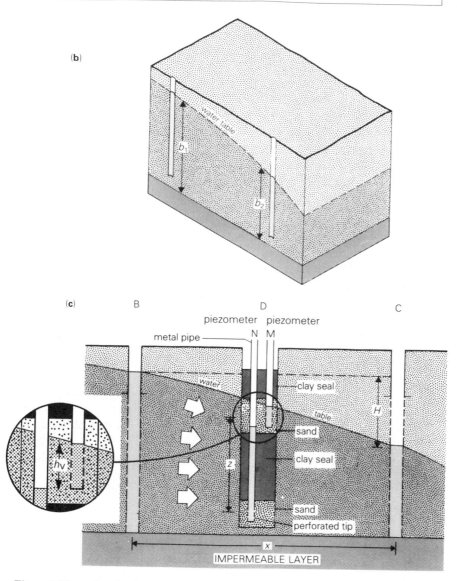

Figure 6.10 continued. (b) In an *unconfined* aquifer, the sloping water table causes a change in the saturated thickness and hence in the area through which flow occurs. (c) In an unconfined aquifer the water table must be sloping if flow is taking place. Because of this slope (shown exaggerated here) there is a vertical component of flow in the upper part of the saturated aquifer which means that a vertical hydraulic gradient must be present in addition to the horizontal hydraulic gradient. The vertical hydraulic gradient can be measured by comparing water levels in the piezometers at M and N, which are open only to a small thickness of the aquifer; the difference in the heads of the two piezometers is h_v, so the vertical hydraulic gradient in this case is h_v/z, where z is the vertical separation of the piezometer tips. The horizontal hydraulic gradient can be measured by comparing water levels in B and C and is H/x.

it is frequently expressed in $(m^3/day)/m$, which reduces to m^2/day. Important British aquifers such as parts of the Chalk and the Permo-Triassic sandstones have transmissivities in excess of $1000\ m^2/day$.

Sometimes the hydraulic gradient H/l is written as I, so Darcy's law may also be expressed as $Q = KAI$ or $Q = TIw$.

In the case of an unconfined aquifer (Figure 6.10b) the slope of the water table is a measure of the hydraulic gradient. Darcy's law still applies, but horizontal flow is occurring only through the saturated part of the aquifer. In this case, the transmissivity is the product of hydraulic conductivity K and b, where b is the *saturated* aquifer thickness. A complication is that because the water table is sloping, b is not constant; in cases like this, either an averaging procedure has to be used before Darcy's law can be applied, or Darcy's law has to be applied to successive 'slices' of the aquifer perpendicular to the flow direction.

Further, because the water table is sloping, the flow is not purely horizontal. As Figure 6.10c shows, there must be a vertical component of flow in the upper part of the saturated portion of the aquifer; this in turn means that there is a vertical hydraulic gradient in addition to the horizontal one. If we observe the water level (which is a measure of the head) in boreholes at B and C we find that there is a difference H, which enables us to calculate the horizontal hydraulic gradient between B and C. If we construct boreholes M and N, at the same horizontal location D, but each open to the aquifer only at a particular and different depth, we find that there is also a difference in water level between them. Boreholes like M and N, which are sealed throughout most of their depth in such a way that they measure the head at a particular depth in the aquifer, are called **piezometers**. The head difference h_v divided by the vertical distance (in this case, z) between the two points at which the heads are measured, represents the **vertical hydraulic gradient** at D.

Because of these vertical hydraulic gradients, the level at which the water stands in a deep borehole may not be exactly the level of the water table. If the flow is downward, the head will decrease with depth, so the water level in a deep well will be below the water table. If the flow is upward, the head will increase with depth and the water level will stand above the water table. Layers of low permeability tend to exaggerate these vertical gradients; many so-called artesian wells – see Chapter 7 – are an extreme example of a condition of upward flow, with the water level in the well standing above ground surface.

In a homogeneous aquifer, the relationship $T = Kb$ is perfectly valid and straightforward. Most real aquifers, however, consist of a combination of layers of varying permeabilities and thicknesses: in this case, strictly, the transmissivity must be worked out by calculating the contribution of each layer – its individual permeability times its thickness – and adding all these contributions together. An 'average' permeability can then be determined by dividing the total transmissivity by the total thickness. The problem

becomes particularly complicated in the case of fissure-flow aquifers like
the Chalk, where the rock mass itself has negligible permeability and
virtually all flow takes place through a few fissures. Here the idea of
'average' permeability is of limited value, since it will vary greatly according
to the rock thickness over which it is averaged.

FLOW TO A WELL

One of the most important applications of Darcy's law is in considering
flow to a well or borehole. Consider a borehole that goes into a confined,
homogeneous, isotropic aquifer of hydraulic conductivity K, and reaches
to the bottom of the aquifer (Figure 6.11a); such a borehole is said to be
fully penetrating. Suppose that the potentiometric surface is initially
horizontal so that the groundwater is not moving at all, in any direction –
an unlikely state of affairs, but it will make the discussion easier! Then, using
a pump inserted into the borehole, we start pumping water out of the
borehole at a rate of Q m³/day.

The action of the pump in withdrawing water from the well causes a
reduction in pressure around its intake, and this in turn creates a head
difference between the water in the borehole and that in the aquifer (Figure
6.11b). Water flows from the aquifer to the borehole to replace that
abstracted, and is in turn drawn up by the pump. Water therefore flows
from further out in the aquifer towards the borehole and so the effect
continues, with the pumping causing a lowering of the potentiometric
surface which spreads outwards from the borehole like a ripple from a stone
dropped in a pond. We eventually reach a situation where the potentiometric surface is being steadily lowered and is sloping smoothly towards the
borehole from all around in the aquifer.

When this occurs, it might seem reasonable to expect that the hydraulic
gradient – the slope of the potentiometric surface – would be constant. This
is not so. Actually, the shape is more like that in Figure 6.11c, with the
hydraulic gradient becoming steeper as the well is approached. To see why
this should be so, consider the situation shown in Figure 6.12.

In this diagram, two imaginary cylinders have been drawn, coaxial with
the borehole. As water is being abstracted from the borehole at Q m³/day,
and as we have an equilibrium condition, Q m³/day must be flowing across
the surfaces of both cylinders. The outer cylinder has radius r_2, so its
circumference is $2\pi r_2$ and the area of its curved surface – the area through
which flow is occurring – is $2\pi r_2 b$, where b is the aquifer thickness. Similarly,
the inner cylinder, radius r_1, has a curved surface of area $2\pi r_1 b$. From
Darcy's law we can therefore write

$$Q = K2\pi r_1 b I_1 = K2\pi r_2 b I_2 \qquad (6.2)$$

where I_1 is the hydraulic gradient at radius r_1 and I_2 the hydraulic gradient

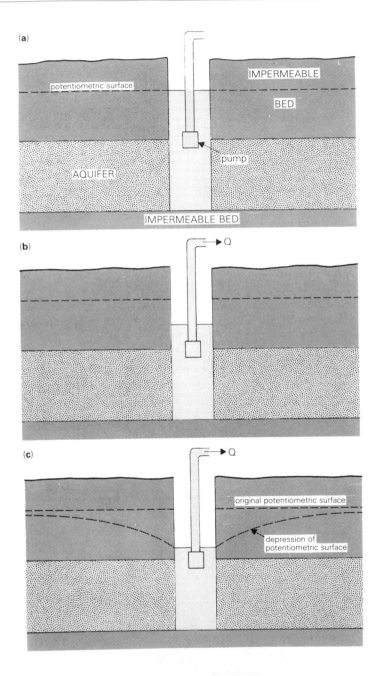

Figure 6.11 The development of the cone of depression. (a) If no flow is taking place, the potentiometric surface is initially horizontal. (b) If the pump is started the head in the well is lowered. (c) This causes water to flow from the aquifer to the well, causing a lowering of the potentiometric surface.

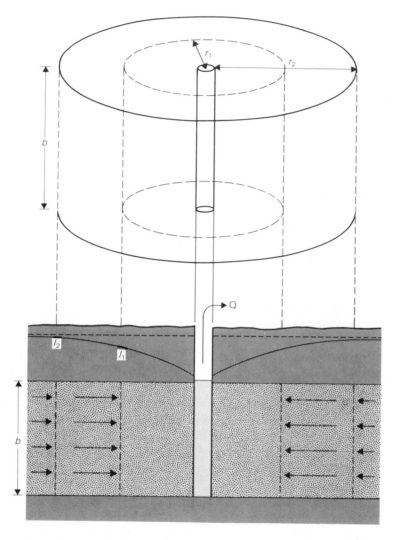

Figure 6.12 Flow to a well. As water flows towards a well, water must flow through successively smaller areas at the same rate. To achieve this the hydraulic gradient must become steeper, causing the characteristic shape of the cone of depression.

at radius r_2. The expression $K2\pi b$ is a constant, so we can see that since r_2 is greater than r_1, I_1 must be greater than I_2 by a proportionate amount in order for the equation to balance and Q to remain unchanged. So this is the main reason why, as the water approaches the borehole, the hydraulic gradient becomes steeper, forming a characteristic lowering of the water table or potentiometric surface called the **cone of depression**.

There can be other reasons too. For the same flow to occur in the same time through a smaller cross-sectional area, not only must the hydraulic gradient become steeper but the actual speed of flow must increase. In some cases the speed may increase to the point at which the kinetic energy, and therefore the dynamic head, is significant. As we saw from Figure 6.7, this increase in kinetic energy must be compensated for by a decrease in potential energy and therefore in static head; the potentiometric surface, which indicates the static head, is therefore lowered further.

A further complication arises in unconfined aquifers. As the water table is lowered in the cone of depression, so the saturated aquifer thickness, b, is reduced. Study of Equation 6.2 shows that if b is reduced, I must again be increased to compensate. But if the hydraulic gradient becomes steeper, the water table in the centre of the cone of depression is lowered more, decreasing b still further so that I must be increased again. Vertical flow components and hydraulic gradients (Figure 6.10c) also come into effect. These effects tend to reinforce each other, and this makes predicting the behaviour of wells in unconfined aquifers more difficult than the behaviour of those in confined aquifers, where the saturated thickness is constant.

Under the influence of the artificially created hydraulic gradient around the borehole, the groundwater percolates through the aquifer until it reaches the immediate vicinity of the well. Here it has one last hurdle to jump – it has to cross the well face and enter the open space of the well bore, before it can travel up or down the well to the pump. By the time it reaches the well face, the water may be travelling at such a speed that it has significant kinetic energy. On entering the well, much of this kinetic energy will be dissipated as turbulence, as the water molecules change direction in their movement towards the pump – movement which requires that some potential energy be converted into kinetic energy. Add to all this the fact that to enter the well the water may have to negotiate some form of slotted lining tube, or a portion of the aquifer whose permeability has been reduced by the act of drilling the borehole, and the head loss involved in this last part of the journey may be a significant part of the total.

This two-part nature of the head loss causes a characteristic steepening of the cone of depression immediately around the well (Figure 6.13). The pumping water level in the well is usually some distance below the depressed potentiometric surface just outside the well face. The total head loss (the distance between the **pumping water level** and the static or **rest water level**) is called the **drawdown**; that part of the drawdown that results from water flowing through the aquifer to the well is called the **aquifer loss**, and that part that occurs as water actually flows into the well across the well face is called the **well loss**. The ratio of aquifer loss to total drawdown is a measure of the **efficiency** of the well as an engineering structure for abstracting groundwater.

These head losses are important, because in order to lift the water to the ground surface, the pump must impart energy to the water. The greater the drawdown, the more energy is needed. This energy is not free; it has to be supplied to the pump in the form of electricity or fuel for the pump motor.

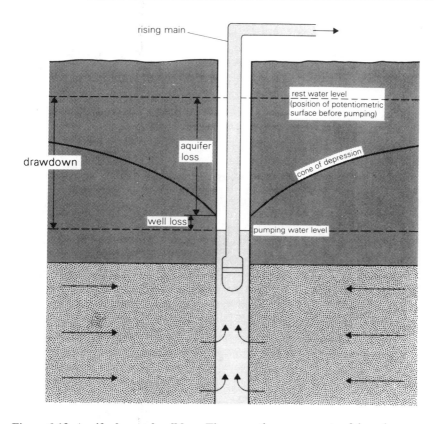

Figure 6.13 Aquifer loss and well loss. These are the components of drawdown.

To save money, therefore, it is essential that the well should be as efficient as possible – essential, in other words, that drawdown should be kept to a minimum.

A useful way of assessing the behaviour of a well is to pump the well at different rates, and measure the drawdown at each rate. The resulting pairs of figures are then plotted in the form of a yield-versus-drawdown graph, often called a **yield-depression curve**.

In a confined aquifer, provided that the pumping water level is above the top of the aquifer (so that the saturated thickness and transmissivity are constant), the yield-depression curve should in theory be a straight line, in accordance with Darcy's law that flow is proportional to hydraulic gradient. In practice, the relationship is more likely to be a curve (Figure 6.14a).

The complication here arises from the fact that a fluid may flow in two different ways. The first way is called **laminar flow**; in this type of flow, the particles of fluid all move smoothly more or less in the same direction as the bulk of the fluid. (By fluid particles here we mean small 'boxes' of fluid, bigger than molecules but small in relation to the passageway through which the fluid is flowing.) In the other type of flow, called **turbulent flow**,

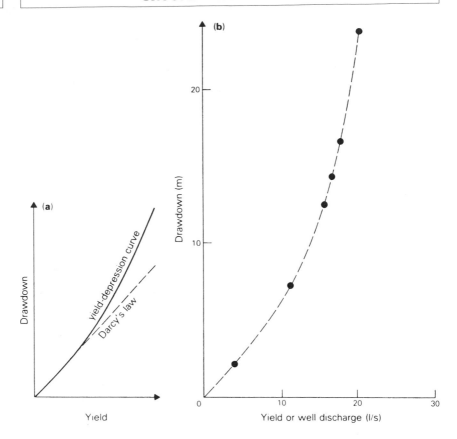

Figure 6.14 Yield-depression curves. (a) Theoretical yield-depression curve for a well in a confined aquifer. If the flow through the aquifer and into the well obeyed Darcy's law, the relationship would give a straight line. But because extra energy is dissipated as the water flows into the well and up or down to the pump, the drawdown is usually greater than that predicted by Darcy's law, and the yield–drawdown relationship follows a curve. (b) Yield-depression curve for a well in the Chalk of England. The marked curvature is a result of the saturated thickness (and hence the transmissivity) of the aquifer being reduced as drawdown increases. (Based on data from the British Geological Survey.)

the motion of the fluid particles at any point can be changing rapidly, both in speed and in direction. In laminar flow, viscous forces are dominant – in other words, the fluid is tending to resist motion of its particles relative to each other. In turbulent flow, the viscous forces are being overcome and the particles are whirling around in a much more unruly fashion. Laminar flow typically occurs when fluid is moving very slowly through small openings (like capillary tubes) or in very thin sheets.

To express the relative importance of viscous forces in any flow condition, a number called the **Reynolds number** (symbol N_R) is used. It is calculated from the formula

$$N_R = \frac{vL}{v}$$

where v is the speed of flow, L is a characteristic length (usually the width of the flow passage or some obstacle in the path of the flow) and v is the kinematic viscosity of the fluid. (All of these parameters must be in consistent units, so N_R has no units – we say that it is *dimensionless*.) In the case of flow in pipes, L is the pipe diameter, and the transition from laminar flow to turbulent flow occurs when the Reynolds number is greater than about 2000.

In flow in a porous medium, an apparent velocity, Q/A, is used for v and the average diameter of the particles of rock (e.g. sand grains) is used for L. The onset of turbulence seems to occur when N_R is more than about 100, but Darcy's law appears to be valid only when N_R is less than 10. However, in most natural groundwater flow conditions, N_R is less than 1.

Darcy's law is therefore valid only for laminar flow, and may become invalid for flow speeds that are at the upper end of the laminar-flow range. At these higher speeds the flow rate may become proportional not to the head difference but to the square root of the head difference, i.e.

$$Q \propto \sqrt{h_L}.$$

Although Darcy's law is applicable to most of the flows of groundwater found in nature, the higher flow speeds that occur immediately around a well frequently result in head losses which increase in this square relationship. This means that if the well discharge Q is doubled, the head loss over this part of the flow path must be increased four times. This explains why the yield-depression curve of Figure 6.14a shows a drawdown that increases faster than the well discharge. It also explains why the designer of a well must do all he or she can to keep the speed of the water as low as possible as it enters the well. The way this is done is described in Chapter 9.

In practice it is difficult to measure directly the extra drawdown which results from well loss. What we can do, by making certain assumptions, is to calculate what the drawdown in the well would be if the only energy losses were those arising from Darcian flow in the aquifer, and compare this theoretical drawdown with that which is actually present; we attribute the difference to well losses, ignoring the fact that some non-Darcian flow may occur in the aquifer. The analysis of the two components of drawdown has resulted in many papers being published, all with the avowed (and admirable) aim of clarifying the problem, but many with the directly opposite result.

Figure 6.14b shows a real example of a yield-depression curve from a well in the English Chalk; here the increasing slope is largely a result of the transmissivity being decreased as the water table was lowered in this (unconfined) aquifer, although turbulence and vertical-flow effects are also believed to have played a part.

SPEED OF GROUNDWATER MOVEMENT

If we know the transmissivity of an aquifer, and the hydraulic gradient, we can use Darcy's law to calculate how much water is flowing
through the aquifer. But suppose we are interested not in the volume
rate of flow – that is, how much water is flowing through, but in the
speed of a water molecule, or of a substance dissolved in the water?
This question may arise, for example, if a potentially harmful substance has entered groundwater and is travelling towards a borehole
(Figure 6.15). How long will it take the pollutant to reach the
borehole?

In discussing laminar and turbulent flow (p. 61) in an aquifer, we
used the idea of an apparent velocity, Q/A. This apparent velocity is
given the symbol q and called the **specific discharge**, and we can then
write Darcy's law as

$$q = KI$$

where I is the hydraulic gradient. Because the hydraulic gradient is
dimensionless, specific discharge has the same units as hydraulic
conductivity, K, namely $(m^3/day)/m^2$, which reduces to m/day. Although these are the same units as speed or velocity, q is not the true
speed of the groundwater. This is because A, the cross-sectional area
that is used to calculate q, is the cross-sectional area of the whole of
the aquifer that we are considering; however, the water is moving
through only part of that aquifer – the pore space.

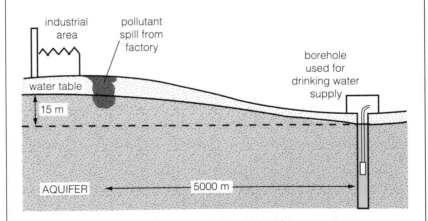

Figure 6.15 Flow of pollutant towards a borehole.

We can see from Figure 6.16a that for the same flow rate to occur through a smaller cross-sectional area, the flow speed must increase. If the flow speed through area A_1 is q when the total area A_1 is available for flow, then the flow speed through the reduced area A_2 will have to be v, where

$$qA_1 = vA_2 = Q.$$

In the simple example of Figure 6.16a the ratio A_2/A_1 is the same as the ratio of volume of the pores to the total volume – in other words it is the same as the porosity. Hence

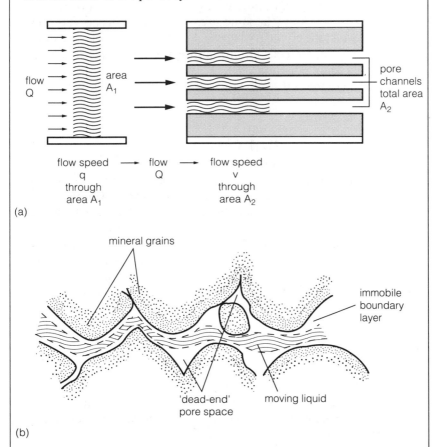

(a)

(b)

Figure 6.16 Flow through pore channels. (a) Water flowing at a volume flow rate of q per unit time per unit cross-sectional area through a pipe has a speed of q units per unit time, and will travel a distance qt in time t. For the same volume rate of flow, q, to occur as the specific discharge through a porous medium, the flow speed must increase to v, because only part (the pore-channel area) of the total cross-sectional area is available for flow. (b) In a natural porous medium, not all of the pore space contributes to water movement.

$$v = \frac{Q}{A_2} = \frac{Q}{nA_1} = \frac{q}{n}. \tag{6.3}$$

Unfortunately, in real aquifers the pores are not neat cylindrical tubes but are much more variable (Figure 6.16b). Furthermore, not all of the pores will be used as flow paths – some of them will be 'dead ends'. Finally, the water will not make use of the full width of the pore channels; there will be an immobile layer – the boundary layer – around the grains. The effect of the variable size of the openings means that we have to talk of a mean flow speed – the flow speed that would occur if the pores, while giving rise to the same porosity and permeability, were all of the same size. The effect of the non-contributory pores and the boundary layer is that the figure for porosity that we need to use in our calculations of flow speed is smaller than that obtained by using Equation 2.1.

The result of this is that our equation for the speed of groundwater movement becomes

$$v_i = \frac{q}{n_d} = \frac{KI}{n_d} \tag{6.4}$$

where v_i is the average speed of groundwater movement and n_d is called the kinematic porosity or **dynamic porosity** – that part of the porosity that is involved in groundwater movement. Although n_d is not easy to measure, it is reasonably close to the specific yield, which can be used as a good approximation for n_d in Equation 6.4.

In the example in Figure 6.15, we can see that the water table falls by 15 m in a distance of 5000 m, so that the average hydraulic gradient, I, is 15/5000 or 0.003. Suppose that the aquifer is a homogeneous sandstone with a hydraulic conductivity of 5 m/day and a dynamic porosity of 0.15 or 15%.

From equation 6.4 we calculate v_i to be

$$v_i = \frac{5 \times 0.003}{0.15} = 0.1 \text{ m/day.}$$

This means that it would take 50 000 days, or more than 135 years, for a water molecule – and therefore presumably for a pollutant – to travel from beneath the industrial site to the borehole. Several points are worth noting:

1. The calculation did not take account of the fact that pumping from the borehole would increase the hydraulic gradient.
2. The assumption that the pollutant moves at the same speed as the water may not be valid. Effects such as diffusion and adsorption (Chapter 13) may mean that the pollutant travels more slowly.

3. The *larger* the dynamic porosity, the *slower* the speed of movement, and vice versa.

The last point becomes very significant when dealing with fissured aquifers. Suppose that the aquifer in Figure 6.15 is a limestone, in which all the permeability and dynamic porosity is contributed by horizontal fissures. The average permeability is still 5 m/day, but the dynamic porosity is only 0.001 or 0.1%. Inserting these values in Equation 6.4 gives

$$v_i = \frac{5 \times 0.003}{0.001} = 15 \text{ m/day}$$

and the water would move from the factory to the borehole in less than one year.

The difference in travel times may not seem significant. If the pollutant is going to reach the borehole eventually, it poses a threat to the water supply from that borehole; does it therefore matter whether it arrives quickly or after a long time? The answer depends to some extent on the nature of the pollutant. Some pollutants may degrade or decompose into harmless by-products if they remain in the aquifer for long enough. Similarly, few harmful micro-organisms will survive in an aquifer for periods in excess of a year. The other side of the argument is that a long travel time may mean that a pollutant is in the aquifer, undetected, for many years, perhaps appearing unexpectedly at a borehole long after the premises that caused the pollution have ceased to operate.

Calculations of this type can be used to define **protection zones** (Chapter 13) around boreholes used for public supply. Within these zones activities that might pollute groundwater can be restricted.

SUMMARY

Groundwater moves from regions where it has high head to regions where it has lower head. **Head** is the height to which the water can raise itself above a reference level (a datum), and is measured in metres; it is a way of measuring how much energy the water possesses. Groundwater usually flows so slowly that its energy due to movement is negligible; the remaining ways in which groundwater can possess energy are by virtue of its elevation and of its pressure. When groundwater moves, some energy is dissipated and therefore a 'head loss' occurs.

When a borehole is drilled into an aquifer, the level at which the water stands in the borehole (measured with reference to a horizontal datum such as sea level) is, for most purposes, the head of water in the aquifer. Except in a few special cases the terms 'head' and 'water level in a borehole' can be regarded as meaning more or less the same thing.

Darcy's law, an empirical law discovered by Henri Darcy in a series of experiments in 1855 and 1856, describes the flow of groundwater through an aquifer. In simple terms, Darcy's law states that the flow rate Q will be directly proportional to the cross-sectional area A through which flow is occurring, and directly proportional to the **hydraulic gradient** I. The hydraulic gradient is the difference in head between two points on the flow path divided by the distance (measured along the flow direction) between them. Thus Darcy's law can be written

$$Q = KAI, \tag{6.5}$$

where K, the constant of proportionality, is called the **hydraulic conductivity**, and depends on the pore geometry of the aquifer and on the viscosity of the water.

The **transmissivity**, T, of an aquifer is the product of the hydraulic conductivity and the saturated-aquifer thickness, b.

If K is constant throughout the aquifer, the aquifer is **homogeneous**; if K is the same in all directions at any point the aquifer is **isotropic**. Darcy's law, expressed in the simple form of Equation 6.5, is valid for aquifers which are homogeneous and isotropic.

When water is pumped from a well, the water level (head) in the well is lowered and a hydraulic gradient is set up towards the well from all around in the aquifer. This causes a **cone of depression** of the potentiometric surface or water table; the reduction in head (lowering of the water level) at the well itself is called the **drawdown**. The drawdown can be considered to consist of two parts: an **aquifer loss**, associated with flow in the aquifer towards the well, and a **well loss**, associated with flow across the well face and into the well.

SELECTED REFERENCES

Darcy, H. 1856. *Les fontaines publiques de la ville de Dijon.* Paris: Victor Dalmont. (See pp. 305–11.)

Fancher, G. 1956. Henry Darcy – engineer and benefactor of mankind. *Journal of Petroleum Technology* **8** (October), 12–14. (A brief biography of Darcy.)

Freeze, R.A. and J.A. Cherry 1979. *Groundwater.* Englewood Cliffs, NJ: Prentice-Hall. (See especially Ch. 2.)

Hubbert, M.K. 1940. The theory of ground-water motion. *Journal of Geology* **48**, 785–944. (A classic and generally very readable account.)

Hubbert, M.K. 1956. Darcy's law and the field equations of the flow of underground fluids. *Transactions of the American Institution of Mining and Metallurgical Engineers* **207**, 222–39.

Todd, D.K. 1980. *Groundwater hydrology*, 2nd edn. New York: Wiley. (See especially Chs 3 and 4.)

Vallentine, H.R. 1967. *Water in the service of man*. London: Penguin. (Contains clear and readable explanations of water movement, viscosity, etc., now out of print.)

7 | More about aquifers

The term 'aquifer' was introduced in Chapter 2. A more formal definition states that an **aquifer** is a geological formation, group of formations, or part of a formation that contains sufficient saturated permeable material to yield significant quantities of water to wells and springs. This definition is by Oscar Meinzer, of the United States Geological Survey. In the case of an unconfined aquifer, it was apparently intended, though not explicitly stated, that the term 'aquifer' should include the *un*saturated part of the permeable material, i.e. the part above the water table, as well as the saturated part. However, we can consider that the *effective aquifer* – the part through which groundwater percolates – extends from the aquifer base to the water table.

PERCHED AQUIFERS

Sometimes a layer of more or less impermeable material occurs above the water table. Infiltrating water is held up by this layer to form a saturated lens, which is usually of limited extent, above the saturated zone of the aquifer proper (Figure 7.1). An occurrence such as this is called a **perched aquifer** (and its upper limit a **perched water table**) because the groundwater in the lens is perched above the saturated zone. Perched aquifers are more common than is often supposed; although they may sometimes be only a few centimetres thick or be present only after a major infiltration event, they may in other cases be several metres thick and extend over large distances. Perched aquifers do not make large or reliable sources of supply, and it sometimes happens that the act of drilling or deepening a well penetrates the impermeable layer and allows most of the perched water to drain away.

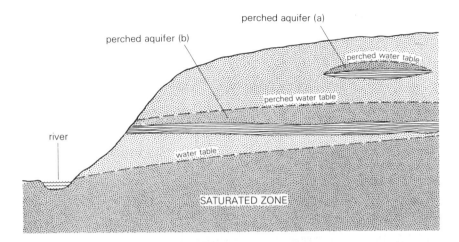

Figure 7.1 Common occurrences of perched aquifers. Perched aquifers: (a) caused by impermeable material of limited extent; (b) occurring where an impermeable bed intersects a valley side some way above the river level. Situation (b) is a more common cause of perched aquifers of moderate extent than is situation (a) – in the latter, the perched aquifer will probably exist only after a period of infiltration.

CONFINED AQUIFERS: THE CONCEPT AND THE MISCONCEPTIONS

In an **unconfined** or **water-table aquifer**, the upper limit of saturation – the water table – is at atmospheric pressure. At any depth below the water table the pressure is greater than atmospheric, and at any point above the water table the pressure is less than atmospheric (Figure 4.4).

In a **confined aquifer**, the effective aquifer thickness extends between the two impermeable layers, and at any point the water pressure is greater than atmospheric. If we drill a borehole through the confining layer, water will rise up the borehole until the column of water in the borehole is long enough to balance the pressure in the aquifer. If we imagine many boreholes drilled into the aquifer, with their water levels joined by an imaginary surface, that surface would indicate the static head in the aquifer. The term **potentiometric surface** was suggested by the US Geological Survey to replace earlier names (such as piezometric surface) for this surface; it can apply to both confined and unconfined aquifers – for example, the water table is a potentiometric surface. (Because the head may vary with depth in an aquifer – for example, as a result of recharge from above causing some vertical flow – each level in the aquifer may have its own potentiometric surface. A borehole that penetrates all or most of the aquifer, and that can receive water from all levels, measures an 'average' head and defines an 'average' potentiometric surface.)

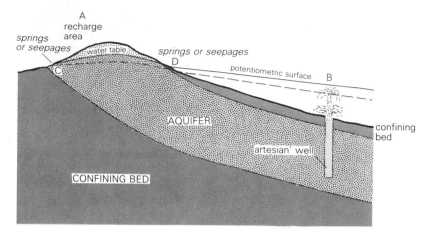

Figure 7.2 The classic explanation of an artesian well.

Confined aquifers are sometimes called artesian aquifers. The term **artesian** was first applied to wells that penetrate aquifers in which the potentiometric surface is above ground level, so that on completion the well overflows or produces water without being pumped (Figure 2.3). The term derives from the Latin name, Artesium, for the Artois region of north-east France where the phenomenon was first studied.

The classic explanation of an artesian well can be seen in Figure 7.2; this type of cross-section used to be common in geography textbooks. It explains the effect in simple 'water finds it own level' terms; rain falling on the aquifer outcrop at A (the **recharge area**) percolates through the aquifer to the well at B. Because of the difference in elevation between A and B, the potentiometric surface at B is above ground level. Rainfall keeps the aquifer 'topped up', the excess water being discharged by springs at C and D. If large numbers of wells are sunk at B, the discharge may exceed the replenishment. Then the potentiometric surface will be lowered (dotted line), the springs at D may cease to flow, and the wells themselves may cease to flow naturally. If the wells are then pumped, the potentiometric surface may be lowered further, below the top of the aquifer, which then ceases to be confined. This state of affairs has arisen in a number of so-called artesian basins, the London Basin – in some areas of which water in the Chalk used to be at sufficient pressure to flow to the surface, but is now as much as 50 m below ground level, and has been as much as 90 m down in the past (Chapter 12) – being one of the best known examples.

The amount by which the potentiometric surface has been lowered below London is sometimes exaggerated, however, and a peculiar misconception has arisen about the fountains in Trafalgar Square. Three wells close to the National Gallery originally supplied water to the fountains, and to the Houses of Parliament and other government buildings. The misconception is that the fountains themselves were originally overflowing wells, with the artesian discharge making pumping unnecessary. The records show however that when the wells were completed in 1847 the water level was already 24 m below sea level. It is doubtful whether overflowing conditions ever existed in this part of London, and the misunderstanding probably arose because of a tendency in the 19th century to describe any well that was drilled, rather than excavated, as 'artesian'.

Confined formations which are exploited as aquifers (as opposed to deeply buried permeable layers which are too deep, or in which the water is too saline, for them to be of economic use) have a recharge area where the formation crops out at the surface. In this area the aquifer is unconfined; the same aquifer is therefore confined in one area and unconfined in another (Figure 7.2), and it should be understood that when we speak of a 'confined aquifer' we mean an 'aquifer where confined conditions exist'; this does not mean that the formation is nowhere unconfined.

The classic explanation of confined aquifers and overflowing wells (Figure 7.2) is reasonably accurate so far as it goes, but it is incomplete. It treats the aquifer as a simple flow conduit, conveying water from the recharge area to the wells that release the water. It ignores storage in the aquifer. It also ignores the fact that all aquifers have recharge and discharge areas which are topographically controlled (Chapter 8): flow in artesian aquifers is in many ways an extreme example of the effects found in unconfined aquifers, with the vertical hydraulic gradients increased as a result of the presence of the overlying confining bed.

It was not until the 1920s that a fuller understanding of the behaviour of confined aquifers began to be achieved. This came about because of a study of the Dakota Sandstone by Oscar Meinzer and Herbert Hard.

The Dakota Sandstone had been exploited as an artesian aquifer in North and South Dakota since 1882. There are actually two aquifers, an upper and a lower, separated by a widespread layer of lower permeability. Meinzer and Hard dealt with the upper aquifer – the one from which most of the wells drew their water – which was overlain by a sequence of shales, varying in thickness from 300 m to 500 m, acting as the confining bed.

When the first wells were drilled into this formation, some of them jetted water to heights of 30 m or more above ground. Pressures and flows were so great that water from some wells was used to drive electrical generators and other machinery. Between 1902 and 1915 there was a marked fall in pressure in the aquifer as more wells were drilled to take advantage of what many people believed to be an everlasting source of water.

By 1915, there were said to be 10 000 artesian wells in South Dakota, and in 1923 it was estimated that there were between 6000 and 8000 artesian wells in North Dakota. In one area for which there were reasonable data, the head declined by an average of nearly 4 m a year between 1902 and 1915.

Meinzer and Hard considered a strip of land, about 10 km north–south and 165 km east–west. They estimated that in the 38 years between 1886 and 1923, the wells in this strip had yielded an average combined flow of about 190 l/s or 16 000 m³/day. They also estimated the rate at which water was percolating eastward through this strip of aquifer from its recharge area in the west. They did this by making assumptions about the thickness and hydraulic conductivity of the sandstone, and from approximate knowledge of the hydraulic gradient. This calculation came out at about 25 l/s or about 2200 m³/day.

It was assumed that this east–west strip of aquifer was like those to the north and south and that all were similarly developed, so that each strip neither gained nor lost to its northerly or southerly neighbours, but was supplied with water from the recharge area to the west of it. On this basis, it was apparent that some extra source of water was contributing to the flow of these wells. The aquifer is underlain and overlain by material of low permeability, so a vertical flow of water into the aquifer was considered out of the question. In the end, Meinzer and Hard concluded that the extra water must somehow have come from storage within the aquifer itself.

ELASTIC STORAGE

At first sight this may seem paradoxical. It is fairly easy to see how water is taken from storage in an *unconfined* aquifer – the water table falls and water drains from the pore space, the amount that drains being governed by the specific yield. But in the case of a confined aquifer, so long as the potentiometric surface is above the top of the aquifer, the pore space is always completely filled with water. How, then, can any water be released from storage?

For answer, think of a motor-car tyre. It is filled with air, under pressure. If you open the valve for a moment, some air comes out and the pressure drops slightly, but the tyre is *still filled* with air. Air is compressible: when we pump it into a tyre we compress it, by forcing the molecules closer together. When we open the valve, some molecules escape and the remainder move further apart. Also, the rubber tyre is elastic; it stretches and increases its volume when we pump air into it, and contracts again when we let air out.

Similar arguments apply to the aquifer. Water is compressible – not as compressible as air, but nonetheless water molecules can be squeezed a little closer together. And, more significantly, aquifers are elastic – the mineral grains of which they are composed can be forced apart slightly by water pressure. Meinzer and Hard concluded that much of the water that had flowed from the wells of North Dakota had come from these sources. As water was taken from the compressible storage, the pressure in the aquifer declined, just as the pressure in the car tyre falls as air is let out.

Ironically, later work on the Dakota Sandstone has suggested that Meinzer and Hard may have been right for the wrong reason. The Dakota Sandstone used to be regarded as a classic 'artesian' aquifer, conveying water from its outcrop areas in the Black Hills and the Rocky Mountain Front Range to the wells in the Dakotas and elsewhere. Subsequent work suggests that the aquifer is more complex, in both its stratigraphy and its hydrogeology, than was once thought. In particular, it seems that significant quantities of water do enter the aquifer through the adjacent beds which Meinzer and Hard regarded as impermeable. Whatever the complexities of the Dakota Sandstone, however, elastic storage in confined aquifers is now an established fact.

The amounts of water that can be stored as a result of these effects are small compared with those held in the pore space, but they can nevertheless be enormous in total. We saw in Chapter 4 that if the water table in an unconfined aquifer falls by a distance z over an area A, then the volume of water that drains from the aquifer is $A \times z \times S_y$ where S_y is the specific yield. We can say that $A \times z \times S_y$ is the volume released from storage. Another way of defining specific yield would therefore be to say that it is the volume of water released from storage, in a vertical column of the aquifer with unit cross-sectional area, for each unit fall in the water table (Figure 7.3a). Alternatively it can be considered as the volume of water taken into storage in the column with each unit rise in the water table.

In the case of a *confined* aquifer, as we have seen, the aquifer remains fully saturated (Figure 7.3b). The weight of the overlying material is supported partly by the solid grains or framework of the aquifer, and partly by the pressure of the water in the aquifer pore space (Figure 7.3c). When water is removed from the aquifer, the water pressure is lowered, and more weight must be taken by the aquifer framework, causing it to compress slightly. The reduction in pressure also causes a slight expansion of the water. To express the combined effect in terms of water released, we define a property called the **storage coefficient**, which is the volume of water released from or taken into storage per unit surface area of the aquifer for each unit change of head (Figure 7.3b). From the wording of this definition, it can apply equally to a confined or an unconfined aquifer.

In the case of an unconfined aquifer, the volume of water that is released from or taken into storage as a result of compressibility effects (of the water and the aquifer framework) is usually negligible in comparison with the water involved in draining or filling of the pore space. For most purposes the storage coefficient of an unconfined aquifer can therefore be regarded as equal to the specific yield.

FLUCTUATIONS OF WATER LEVEL

Artesian wells, spewing water to a considerable height above the ground, have an undoubted fascination. Largely because of their attraction to the authors of geography textbooks, overflowing wells are probably second only to limestone caverns as the aspect of groundwater with which most

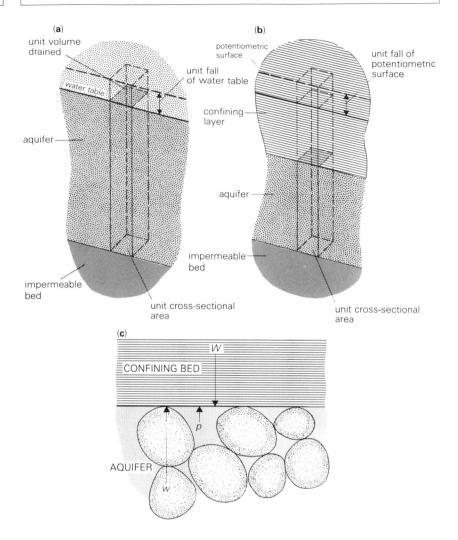

Figure 7.3 Storage concepts. (a) The specific yield in an unconfined aquifer. (b) The storage coefficient in a confined aquifer. (c) The pressure (weight per unit area) W of the overlying strata is balanced partly by the pore water pressure p (measured relative to atmospheric pressure) and partly by the aquifer framework, whose *average* grain contact pressure is w, i.e. $W = w + p$. If p is reduced (e.g. by abstraction of water) then w must increase to compensate (the aquifer framework must take a greater share of the load) and the grains are pressed closer together.

people are likely to be familiar. There are other aspects of the behaviour of confined aquifers which, though less spectacular, are equally fascinating. One of these aspects is the response of wells in confined aquifers to changes in atmospheric pressure.

Comparisons of records of water levels in such wells with records of atmospheric pressure from nearby barometers or barographs show that as pressure rises, water level falls, and vice versa. The effect is not seen in unconfined aquifers. The explanation is apparent from Figure 7.4. In the case of an unconfined aquifer (Figure 7.4a), the water table (by definition) is at atmospheric pressure. Any change d in atmospheric pressure P_a can be transmitted directly to the water table both in the aquifer and in a well, and so the heads remain equal and no measurable change in water level occurs. In the case of a confined aquifer (Figure 7.4b), the pressure W of the

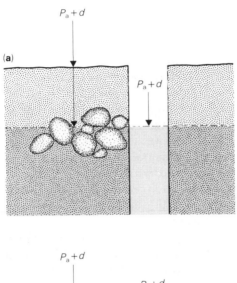

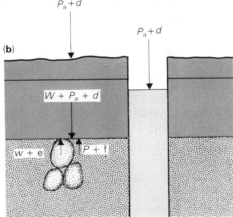

Figure 7.4 Barometric effects. (a) In an unconfined aquifer any increase d in atmospheric pressure P_a is transmitted equally to water in the aquifer and to water in a well. (b) In a confined aquifer, some of the increase (e) is taken by the aquifer grains, and the remainder (f) is taken by the pore water. So, before the change in pressure, we can say that $W + P_a = w + P$. After the change, we have $W + P_a + d = w + e + P + f$, which means that the increase f in the pore-water pressure is less than d. In the well however the full increase d is transmitted to the water surface, forcing some water from the well into the aquifer.

overlying confining bed and the pressure P_a of the atmosphere are carried partly by the aquifer framework, and partly by the water, which is of course at a pressure P greater than atmospheric. Any increase d in atmospheric pressure is transmitted to the top of the aquifer, and is similarly shared between the framework and the water; where a well is present, however, the pressure increase is transmitted directly to the water. The water in the well is now under more pressure (and therefore at a greater head) than the water in the aquifer; therefore some water flows from the well into the aquifer, depressing the water level in the well, until the heads of water in the well and aquifer are again equal. Such a change in water level is termed a **barometric fluctuation**. A fall in atmospheric pressure produces the opposite effect.

If the aquifer is rigid, consolidated rock, most of the increase in atmospheric pressure (transmitted as extra weight by the confining layer) will be borne by the aquifer framework, and the pressure of water in the aquifer (the pore-water pressure) will scarcely change. There will therefore be a relatively large difference between the pore-water pressure and the pressure in the well. Hence the barometric fluctuation will be much larger than in the case of an unconsolidated, weak, aquifer, where most of the increase in pressure will be transmitted to the pore water instead of being taken up by the aquifer framework, resulting in little difference in pressures between the water in the well and the pore water.

Atmospheric pressure changes are usually measured in millibars or in pascals, but they can be expressed in terms of head of a column of water and measured, say, in metres. Then the ratio of the change in water level in a well to the change in atmospheric pressure is called the **barometric efficiency** of the aquifer. Barometric efficiencies are often expressed as percentages. For the reasons explained above, rigid aquifers have high barometric efficiencies – perhaps as high as 70 or 80% – while unconsolidated aquifers usually exhibit much lower values. Unconfined aquifers should have barometric efficiencies of zero, but there are examples of such aquifers apparently exhibiting significant barometric fluctuations. Such a fluctuation occurs when for some reason the full change in the atmospheric pressure is not transmitted directly to the pore water in the aquifer – in other words, the aquifer is not behaving as a true water-table aquifer.

Any event that causes a change in the pressure of the pore water in a confined aquifer will result in a change in water level in wells that penetrate that aquifer. This is because, at any level in the aquifer, a change in water pressure implies a change in head. This change in head, transmitted through the aquifer to the well, manifests itself as a change in water level in the well, which – as we have observed before – is essentially a manometer.

A great variety of events can and do cause pressure changes in confined aquifers. Railway trains can exert a sufficient load on an underlying confined aquifer to cause significant changes in water levels in wells. Major earthquakes can produce effects in wells several thousand kilometres away. The rise and fall of sea level during a normal tidal cycle will exert a loading and unloading effect on a confined aquifer which extends beneath the sea; the resulting changes in aquifer pore-water pressure may be observed in wells several kilometres from the shore.

The study of such water-level fluctuations can reveal important information about the aquifer. There is little doubt however that in general the most important fluctuations in groundwater level result from changes in the amount of water in storage in the aquifer. The principle can be understood by thinking of the aquifer storage in terms of a bucket. If we add water to the store, or bucket, the water level rises; if we remove water, the level falls. If we add more water than the bucket can hold, it overflows: similarly, when the water table rises above a certain level, groundwater will leave the aquifer in the form of springs or as seepage to rivers and streams (see Chapter 8).

One of the differences between the aquifer and our bucket is that the entire volume of the bucket is available to hold water, whereas most of the aquifer consists of mineral grains and water held in place by capillary forces (the specific retention); the bucket, unlike the aquifer, has a 'specific yield' of 100%. Suppose the bucket to be a cylindrical and unusually large one, with a base area of 1 m^2 (Figure 7.5a). If we pour in 100 litres of water then, since 100 litres is 0.1 m^3, the water level in the bucket will rise by 0.1 m. If we now take an identical bucket and fill it with large cobbles, arranged in such a way that they occupy 90% of the volume (leaving only 10% porosity), and then add 0.1 m^3 of water, the water level in the bucket will rise by 1 m

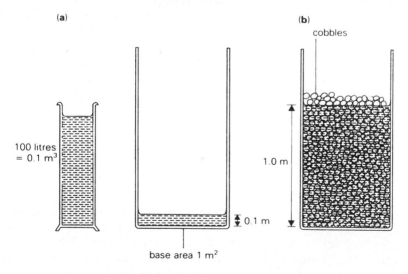

Figure 7.5 The influence of storage on water-level fluctuations. When 100 litres (0.1 m^3) of water is poured into a cylindrical bucket of base area 1 m^2, the water level rises by 0.1 m. When the same bucket is filled with cobbles to give a porosity of 10%, the same volume of water causes a water-level rise of 1 m.

(Figure 7.5b). This is because, for 0.1 m³ of storage space to be filled, 1 m³ of our cobble 'aquifer' has to be affected. If instead of coarse cobbles we filled the bucket with fine sand, with a specific yield of only 1%, then the same volume of water would produce a water-level rise of 10 m. (We should need a very tall bucket!)

If 1 mm of **effective rainfall** (that is, rainfall in excess of evapotranspiration) falls over an area, and if no infiltration or runoff occurs, then the rainfall will result in a layer of water 1 mm deep over the ground surface. If this layer of water then infiltrates into an aquifer with a specific yield of 0.10 (or, in percentage terms, 10%), then the water table will rise by 10 mm; if the specific yield were 0.01, or 1%, the water table would rise by 100 mm. In general, we see that:

$$\text{rise in water table} = \frac{\text{infiltration}}{\text{specific yield (expressed as a fraction)}}.$$

Typically, in an aquifer with a low specific yield, the water table rises a long way in response to recharge and, in the same way, falls a long way in response to withdrawal of water from the aquifer. Aquifers with high specific yields generally show small water-table fluctuations. Measuring the rise in water table in response to a given amount of infiltration therefore provides a means of estimating the specific yield.

Such estimates must be treated with caution, however. In the first place, they provide a measure of specific yield only for the zone of fluctuation; the properties of the aquifer may be different beneath this zone, so that if the water table were lowered further, perhaps by pumping, a different value of specific yield might be applicable. Second, it is rare for infiltration to occur uniformly over an area, or for the recharge to move down to arrive simultaneously at a horizontal water table. Usually, recharge will be un-even, in terms of both time and space. A large amount of rainfall may occur over one part of the aquifer during a storm; the resulting recharge will cause a temporary rise in the water table beneath that area. Immediately this rise begins, groundwater movement will take place away from that locality, so that the size of the final fluctuation in water level will depend not only on the specific yield but on the ease with which the water can be redistributed – in other words, on the transmissivity.

In the case of a confined aquifer, recharge generally occurs as a result of infiltration over an area where the aquifer is unconfined (A in Figure 7.2). This causes a change in the position of the water table in the recharge area, and this is transmitted as a pressure change to the confined part of the aquifer, to cause a corresponding change in the position of the potentiometric surface. Withdrawal of water from the confined part of the aquifer has a direct effect. The storage coefficient of a confined aquifer is generally orders of magnitude lower than the specific yield of a water-table aquifer (a typical storage coefficient for a 30-m thick aquifer would be about 10^{-4} or 0.01%, so large changes in potentiometric level can result from the withdrawal of water. Balancing this, however, is the fact that confined aquifers often extend over large areas, and the large total amount of water stored in such an aquifer may support abstraction for a long time.

Earlier in this chapter, we compared a confined aquifer to a motor-car tyre. A car tyre stretches and expands slightly when inflated, and contracts correspondingly when deflated, and it was implied that a confined aquifer does the same. If this is so then the withdrawal of large volumes of water from confined aquifers at rates much greater than natural replenishment should lead to a contraction of the aquifers and a corresponding subsidence of the land surface. The occurrence of such subsidence is powerful evidence for the elastic behaviour of confined aquifers. Notable examples of such subsidence have occurred in the San Joaquin Valley of California, around Shanghai, in southern Taiwan, and around Mexico City; in all cases, the ground surface has subsided by several metres. In Mexico City, in particular, considerable damage has occurred to buildings. Pumping from aquifers beneath Venice produced subsidence of about 0.2 m which, given the location of the city and other factors, has proved to be very serious.

Such spectacular effects, although serious, are comparatively rare; they tend to occur in thick deposits containing fine sands, silts and clays, and in many cases are the result of excessive pumping from aquifers. Unfortunately, it is often assumed that subsidence is an inevitable consequence of groundwater abstraction, and local people sometimes oppose the development of groundwater-abstraction schemes under the impression that these will lead to subsidence on the same scale as that associated with coal mining. Such fears often stem from false concepts of groundwater occurrence, such as the ideas of underground rivers and lakes mentioned in Chapter 2.

ROCK TYPES AS AQUIFERS

Several factors can provide a basis for classifying rocks – for example chemical composition, mineral composition, age and texture – but, as a starting point, it is usual to group all rocks into three main types, depending on their origin. These are *sedimentary*, *igneous* and *metamorphic*, and most of the world's major aquifers are of sedimentary origin. Igneous and metamorphic rocks, in general, are far less important as sources of groundwater. To see why, it is only necessary to consider how the three rock types form.

Igneous rocks

Igneous rocks form by the cooling and solidification of molten rock or **magma**, which may be forced into other rocks, forming an **igneous intrusion**, or which may be **extruded** at the surface (as for example in a volcanic eruption). This method of formation means that there are usually few voids in the rock at the time of its formation – perhaps just a few, small, unconnected cavities or **vesicles** caused by the presence of bubbles of gas. The rock therefore has very little initial porosity or permeability; only in the case of extrusive igneous rocks (lava) are there likely to be large or interconnected openings (pp. 92, 95–6). Basalt lavas, in particular, are well known for their tendency to form columns separated by sizeable cracks or joints as the lava cools and contracts.

Later in the life of a rock, porosity and permeability may increase. Weathering may weaken and remove some minerals to create voids or to open up joints; tension resulting from movements in the Earth's crust, or stress release as the weight of overlying rock is removed by erosion, may cause fractures to develop and open. To distinguish between voids present when the rock was formed and those which develop as a result of later processes, some workers use the terms **primary porosity** and **secondary porosity**.

Metamorphic rocks

Metamorphic rocks are formed by the alteration of other rocks under the action of heat or pressure. Small occurrences may be due to the baking of other rocks by hot magma, in which case existing porosity may be preserved, but large occurrences of metamorphic rocks are the result of processes deep within the Earth's crust. The temperatures and pressures involved mean that these rocks are altered and compressed to such an extent that voids are destroyed. Only when these rocks are brought back to the surface, as overlying rocks are removed by erosion, is there a chance of secondary porosity developing in the same way as for igneous rocks.

Sedimentary rocks

Sedimentary rocks form as a result of deposition of particles, which are often derived from the weathering and erosion of other rocks. This deposition usually takes place under water, frequently on the sea bed, but it may occur in river beds or lakes, or even on dry land. The nature of the process means that particles will be deposited with spaces between them; the size of these voids will depend on the sizes of the particles, and on how well sorted they are. Clearly, small grains will have small pores between them; large grains, all of one size, will have large pores. If the sediment contains large and small grains – i.e. if it is 'poorly sorted' – the small grains will tend to occupy the voids between the larger ones, leading to a lower porosity than in the case of well-sorted sediments.

Fine sediments, such as clays and silts or fine sandstones, may have high porosities, but the pores are so small that surface tension or molecular forces prevent water movement, so that in these materials permeability is low. Coarse sands and gravels, especially if they are well sorted, are very permeable.

After deposition, water percolating through the sediment may deposit material brought in solution from elsewhere. This deposition on and around the mineral grains is termed 'cement', and it binds the sediment together. The process may be assisted by compaction resulting from burial beneath other layers of sediment. In this way a **non-indurated sediment** or **unconsolidated sediment** becomes an **indurated sediment** or **consolidated sediment**. At the same time, porosity is reduced; in extreme cases the cement may fill up almost the whole of the primary porosity.

Once a sediment has become consolidated it can be subject to fracturing and the development of secondary porosity in the same way as igneous or metamorphic rocks.

UK AQUIFERS

The United Kingdom has a varied and interesting geology, a fact that explains the diversity of scenery within a small area. In Britain, in very general terms, older rocks occur in the north and west and younger rocks in the south and east (Figure 7.6). The older rocks which form most of Scotland, Wales and Northern Ireland are mainly metamorphic and igneous rocks or well-indurated sedimentary rocks which have low permeabilities, so these areas have little in the way of useful aquifers. In Scotland and Northern Ireland, groundwater is used to provide about 5% of public water supplies, while in England and Wales, taken together, about one-third of the water used for public supply is provided by groundwater.

Total groundwater abstraction in England and Wales is about 2400 million cubic metres (or 2.4×10^{12} litres) per year. That is enough water in a year to cover the whole of England and Wales to a depth of about 15 mm, or to supply every man, woman and child in England and Wales with 49 000 litres of water each year (or 135 litres/day).

The main aquifers in England and Wales consist of partially indurated sedimentary rocks. In terms of the quantity of water abstracted, the Chalk is the most important, followed by sandstones of the Permian and Triassic Systems. Abstractions from the Chalk account for about half the total groundwater abstracted in England and Wales, and those from the Permo-Triassic sandstones for about a quarter. These two aquifer groups therefore deserve consideration in some detail.

Chalk

Chalk is a soft white limestone. Limestones are composed of calcium carbonate, and they usually consist of the skeletal remains of aquatic organisms – shell fragments are a common source. The Chalk of Britain

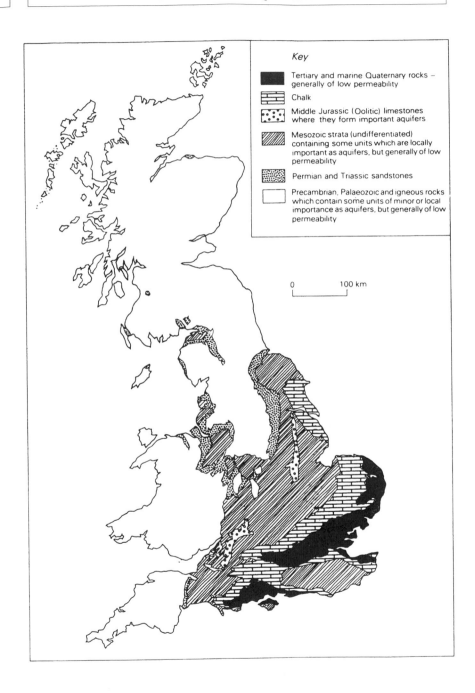

Figure 7.6 Simplified geological map of Great Britain, showing the most important aquifer groups. (Minor outcrops omitted.)

was formed in the late Cretaceous period of geological time, and consists of shell fragments and foraminifera with sizes typically between 10^{-2} and 10^{-1} mm (10 to 100 μm), set in a finer matrix. The matrix is so fine that it was at one time believed to be of inorganic origin, perhaps a chemical precipitate. It was not until the availability of the electron microscope in the 1950s that this finer material was also shown conclusively to be of organic origin, consisting of whole or broken minute calcareous shells (coccoliths) of plankton (Figure 2.1b).

At the time of deposition, the Chalk probably had a porosity of at least 70%, the high value being partly due to the hollow shells. In southern England there has been limited cementation, and porosities of 40% are still common throughout much of the aquifer; in Yorkshire and Lincolnshire, cementation has been more extensive, reducing porosities to 10 to 20%. In Northern Ireland, the process has gone further; in the Chalk here, known as the 'White Limestone', porosities of less than 5% are common.

Most of the Chalk is a pure white limestone containing lumps or layers of flint, but in the lower part of the formation clay minerals and marl bands are common, imparting a greyish colour. Flints are absent from this lower part.

The fine-grained nature of the Chalk means that the pores and pore necks are correspondingly small, the latter typically less than 1 μm. Most pore water is therefore held virtually immobile by capillary forces: in spite of the high porosity, specific yield is therefore low – generally about 1%. Permeability is also low if measurements are made on small samples, which take into account only the intergranular permeability. (The example in Figure 2.1b, from a borehole on the Berkshire/Hampshire border, has a measured porosity of 46% and a hydraulic conductivity of 0.004 m/day.) Typical hydraulic conductivity values measured in the laboratory on Chalk samples are between 10^{-3} m/day and 10^{-2} m/day; given representative thicknesses for the Chalk of 200 m to 500 m, this would imply transmissivities of less than 5 m²/day. In practice, measurements of transmissivity from wells in the Chalk give values that frequently exceed 1000 m²/day. The difference is caused by the presence of cracks or fissures; these are usually in three directions, often more or less mutually perpendicular, one set being approximately parallel to the bedding. These have quite limited openings (usually less than 1 mm) except where they have been enlarged by the dissolution of calcium carbonate, when they may have openings of several millimetres (Figure 7.7); occasionally there are openings more than a metre in height. In general, the enlargement of fissures by dissolution has occurred beneath valleys much more than it has beneath hills. As a result, wells in valleys are usually much more successful than those sunk into chalk in upland areas.

The Chalk is thus an unusual aquifer. Fissures may contribute about 1% porosity, but almost the whole of the specific yield and more than 99% of the transmissivity. Intergranular pore space contributes a porosity of 20 to 50% of bulk volume, but very little to specific yield or transmissivity. At

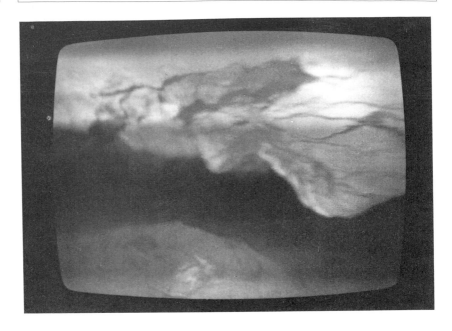

Figure 7.7 Fissure permeability in the Chalk. Photograph taken using a closed-circuit television camera of the wall of a Chalk borehole, showing a fissure that has been enlarged by solution. The view is of an area about 80 mm × 60 mm. (Photograph published by permission of the Director, British Geological Survey.)

great depths, fissures tend to be closed by pressure. Further, at significant depths below sea level there is little or no circulation of fresh water to enlarge fissures by dissolution. In most areas, therefore, it is the upper part of the Chalk that is the effective aquifer. Under these circumstances and with perhaps one or two fissures contributing 80% of the transmissivity, formulae like

$$T = Kb$$

(p. 56) cease to be of practical significance.

Because of the high transmissivity of the Chalk, hydraulic gradients are usually low and the water table is correspondingly flat. Thus in areas with marked relief the water table is often far below ground level, giving rise to an unsaturated zone that may in places reach 100 m in thickness. The small pore sizes and corresponding high specific retention mean that this unsaturated zone is almost fully saturated – only fissures and perhaps a few large pores drain under gravity. However, the pore sizes are such that a great deal of the water can be removed by plants, and there seems little doubt that in dry weather water can move up in limited quantities from depths of a few metres in response to evapotranspiration losses.

Permo-Triassic sandstones

In contrast to the Chalk, which is a marine deposit, the sandstones of the Permian and Triassic systems in Britain were laid down at a time when most of the country was desert or semi-desert. Some of the sandstones appear to have originated as sand dunes, which have subsequently become partially cemented; the majority were deposited by apparently ephemeral rivers and lakes in a semi-arid environment. Deserts are not renowned for an abundance of vegetation or wildlife, and hence organic remains – the fossils that geologists use as a major means of comparing the ages of rocks – are rare in these sandstones. It is therefore difficult to decide whether some of the deposits are Permian or Triassic, and they are often grouped together as Permo-Triassic.

In addition to sandstones, the Permo-Triassic system includes conglomerates, siltstones and mudstones. The nature of their deposition means that different materials were being deposited in different places at the same time, so that there is a great deal of lateral variation in the rocks.

In general, the Permo-Triassic sandstones form a much more 'normal' aquifer than the Chalk. Porosity depends on the degree of sorting and rounding, the packing and cementation and so is highly variable, but values of 20 to 35% are usual; specific yield, which is controlled by grain size as well as the other factors, is typically 15 to 25%, although in practice variations in pore size throughout the aquifer may inhibit drainage of the pore space. Hydraulic conductivity values are typically between 1 m/day and 10 m/day in the coarser sandstones, and between 10^{-1} m/day and 1 m/day in the finer deposits. Desert sandstones, deposited as sand dunes, generally contain rounded grains (Figure 2.1a) which have been rolled by wind action; because finer material is blown away, they are also well sorted. These round grains cannot pack tightly, so such sandstones (e.g. the Penrith Sandstone of the Eden Valley) form some of the most permeable of British aquifers, with hydraulic conductivity values as high as 20 m/day. (The example in Figure 2.1a, from the Permian sandstone of the Eden Valley, Cumbria, has a porosity of 31% and a hydraulic conductivity of 7.7 m/day.)

There is evidence that fissures are important locally in these sandstones and may play a major part in allowing water to flow easily into wells, but they may not carry water on a regional scale in the same way as in the Chalk. Because the intergranular permeability is relatively high, it is easier for water to move from intergranular pore space into fissures, or vice versa, than is the case in the Chalk. The relatively high specific yield means that water-table fluctuations are usually much smaller than in the Chalk.

NORTH AMERICAN AQUIFERS

The North American continent contains some of the most heavily developed and most extensively studied aquifers on Earth. The dominant use of
groundwater in North America is for irrigation; in the United States, this
accounts for about two-thirds of all groundwater used. The importance of
groundwater varies greatly between regions. In Canada, which has enormous supplies of surface water, groundwater accounts for only about 10%
of public supplies; in much of Mexico and in parts of the United States, it
is almost the sole source of water.

Within this large continental area groundwater occurs in, and is extracted from, many types of rock ranging in age from Precambrian to
Recent. The hydrogeology of many of these formations has been described,
often in publications of the national and state surveys of the relevant
countries; the work of the US Geological Survey deserves special mention.
An excellent summary of the hydrogeology of the continent is provided by
the *Hydrogeology* volume of the series on the Decade of North American
Geology, edited by Back, Rosenshein and Seaber (1988); shorter summaries
of some regions are available in the book by Driscoll (1986). Figures 7.8,
7.9 and 7.10 are simplified maps showing the main hydrogeological features
of the continental United States, Canada, and Central America. The following paragraphs summarize key points of some of the regions where groundwater is of most importance (Figure 12.3), and of the aquifers that supply
that groundwater.

The High Plains aquifer

The High Plains aquifer (Figure 7.8) underlies an area of about 450 000 km^2
in the central part of the United States. It extends from South Dakota and
Wyoming in the north to New Mexico and Texas in the south. The aquifer
consists mainly of unconsolidated deposits of Tertiary and Quaternary age,
principally sands and gravels laid down by river systems. The most important unit is the Ogallala Formation, a late Tertiary deposit with a maximum
thickness of about 215 m. In Nebraska the Ogallala Formation is overlain
by an extensive deposit of wind-blown sands covering an area of about
52 000 km^2 and forming the famous Sand Hills.

Because the High Plains is essentially a plateau, groundwater drains
easily from the permeable deposits, the slope of the water table being
generally eastwards. Consequently the saturated thickness averages only
about 60 m, though it can vary from zero to as much as 300 m. The area is
one of the most important agricultural areas in the United States. It is
estimated that in 1980 about 170 000 wells pumped about 2200 million m^3
to irrigate 54 000 km^2. Abstraction is made relatively easy by transmissivities which can be as high as 10 000 m^2/day, with yields of individual wells
ranging from 6 litres/second (l/s) to more than 150 l/s. The climate is of the

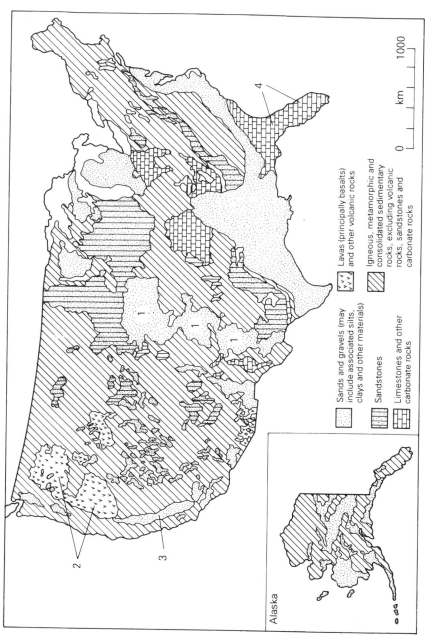

Figure 7.8 Simplified geological map of the continental United States. The rock types are grouped according to their hydrogeological characteristics, with superficial deposits omitted. The most important aquifers, described in the text, are numbered as follows: 1. High Plains aquifer; 2. Columbia Plateau Basalts; 3. Central Valley of California; 4. Floridan and Biscayne aquifers. (Copyright, R.C. Heath, *Journal of Groundwater*. All rights reserved.)

Legend:

- Sands and gravels (may include associated silts, clays and other materials)
- Sandstones
- Limestones and other carbonate rocks
- Lavas (principally basalts) and other volcanic rocks
- Igneous, metamorphic and consolidated sedimentary rocks, excluding volcanic rocks, sandstones and carbonate rocks

Alaska

0   km   1000

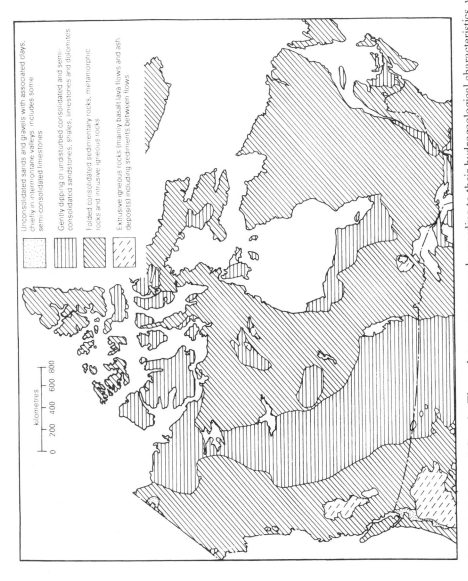

Figure 7.9 Simplified geological map of Canada. The rock types are grouped according to their hydrogeological characteristics, with superficial deposits omitted. From the map compiled by R.C. Heath in Back, Rosenshein and Seaber (1988).

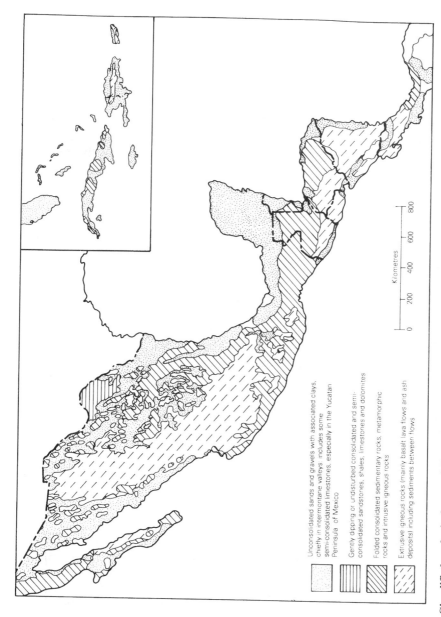

Figure 7.10 Simplified geological map of Central America. The rock types are grouped according to their hydrogeological characteristics, with superficial deposits omitted. Inset shows southern Florida and the Greater Antilles. From the map compiled by R.C. Heath in Back, Rosenshein and Seaber (1988).

dry continental type, with potential evapotranspiration generally exceeding precipitation for much of the year over much of the area. Recharge is therefore limited, except in the north and from the beds of ephemeral streams (Chapter 8).

Columbia Plateau Basalts

A large area in the north-west of the United States is underlain by volcanic rocks – principally basaltic lava flows – interbedded with or overlain by alluvium and lake sediments (Figure 7.8). These rocks form what is in general a high plateau region, drained mainly by the Columbia River and its tributaries, the most important tributary being the Snake River. The rocks range in age from Miocene to Recent, some of them having been deposited within the last 1000 years. In the central part of the area the thickness of these rocks, built up from many successive flows, is known to exceed 3000 m. Individual flows range from less than a metre to more than one-hundred metres in thickness.

The bulk of horizontal groundwater movement takes place through a relatively small proportion of the total thickness. After eruption, the top surface of a flow cools and begins to solidify much more quickly than the central part. The crust on the upper surface may therefore be broken by continuing internal movement of lava. This broken surface also becomes weathered and sediments may accumulate on top of it, the factors combining to give a unit of significant permeability. The base of the succeeding flow is usually riddled with small openings called vesicles, formed by bubbles of gas or steam, together with larger openings where vegetation may have temporarily formed a mould around which the lava cooled. The combined tops and bases of successive flows form thin but often highly permeable layers which behave as individual aquifers, separated by the less permeable bulk of each flow which acts as a confining layer.

These interflow zones are often so permeable that a well need not penetrate more than about 50–100 m below the water table to obtain adequate yields. This means that there are relatively few deep boreholes to provide information. The studies that have been made suggest that these lava sequences are among the most productive aquifers on Earth, with transmissivities ranging from 2000 to more than 400 000 m^2/day and yields from individual wells as high as 450 l/s. There are however large variations. Storage coefficients are generally no more than 1%, but may be as high as 7% in the Snake River basalts.

The Columbia Plateau contains deep valleys that have been used as sites for reservoirs. Surface water has been used for irrigation, and excess irrigation has added to natural recharge of the aquifer; in places this has had the effect of leading to greatly increased flows from springs.

The Central Valley of California

The Central Valley of California occupies an area about 650 km long and 80 km wide forming a trough between the Coast Ranges and the Sierra Nevada. It consists of the valleys of two major rivers, the Sacramento in the north and San Joaquin in the south, which meet in a large deltaic area and drain to San Francisco Bay. The valley is a structural trough formed where two of the major plates of the Earth's crust have collided; the trough is filled by sediments ranging in age from Jurassic to Recent. These sediments are as much as 15 000 m thick in the Sacramento Valley. The earliest deposits are marine, but the later ones were laid down by rivers and lakes as the trough became filled. The later deposits consist mainly of unconsolidated sands, gravels, silts and clays, though volcanic rocks are also present.

The rocks have variable properties. Although permeable units are inter-bedded with less permeable layers, the whole sequence behaves more or less as a single unit, with abstraction from one part eventually affecting adjacent areas. The more permeable units have hydraulic conductivities ranging from 30 to 600 m/day, with transmissivities as high as 10 000 m^2/day recorded. Wells sunk into these coarser deposits commonly yield 30–60 l/s.

The marine deposits contain saline water. Above this zone fresh water is present to a depth of about 300 m and in places to more than 1000 m. This fresh water supplies about 40% of the total water use in the San Joaquin Valley in normal years, and a much higher proportion in drought years. Almost all of the abstracted groundwater is used for irrigated agriculture. In the early years of its development, the groundwater of the Central Valley was thought to be an almost inexhaustible supply of water, and was abstracted with little regard for the consequences. One of the consequences has been the widespread subsidence mentioned earlier in this chapter; another has been the ingress of sea water (Chapter 13).

The Floridan and Biscayne aquifers

The Floridan aquifer system underlies Florida, south-eastern Georgia and the adjacent area of Alabama. It consists of a thick sequence of limestones, mainly of Tertiary age, that are believed to have been deposited in shallow seas similar to those around the Bahamas at present. The limestones form a productive aquifer that is confined over much of the area. Where it is unconfined, or where the confining layer is relatively thin, the natural circulation of water has increased the permeability by dissolving the lime-stone and creating large openings; in these areas, transmissivities may be as high as 100 000 m^2/day. Even in the confined areas, transmissivities are frequently in excess of 1000 m^2/day.

In south-eastern Florida, the confining layer above the Floridan aquifer is in turn overlain by another sequence of limestones called the Biscayne aquifer, which is the main source of groundwater for much of south-eastern

Florida. Like the Floridan, the Biscayne aquifer owes much of its permeability to the presence of solution features; in places the transmissivity approaches 200 000 m²/day. Both the aquifers are at risk from ingress of sea water, but the danger is greater for the Biscayne aquifer.

OTHER AQUIFERS – UNCONSOLIDATED SEDIMENTS

In the United Kingdom, unconsolidated sediments are of minor importance as aquifers, but in other parts of the world they may be the main or only source of groundwater and they provide the bulk of the world's developed aquifers. They are generally of relatively recent origin, and lack of compaction and cementation means that they include some of the most permeable natural materials. Of particular importance are sands and gravels in river valleys, which are usually very permeable but of limited thickness and extent unless the river is a major one; the small size of UK rivers means that alluvial aquifers are of only local importance.

Where the valley is controlled by geological faulting, so that the deposits are filling a trough of structural as well as erosional origin, the deposits may reach thicknesses of hundreds or even thousands of metres; the Central Valley of California (see above) is an example. In other cases the deposition of sediment takes place over a larger area but with smaller thicknesses. The groundwater of these broad valleys is generally a more important resource than that of the deeper fault-controlled valleys; not only do the broader valleys contain a greater total volume of sediment, but the water is frequently of better quality, as the very deep deposits of the rift-type of valley may contain saline water. The groundwater of the Ganges delta, for example, is vital to the peoples of India and Bangladesh.

Unconsolidated sediments were probably the first aquifers to be developed. River alluvium, in particular, was probably an obvious choice for early wells, offering ease of excavation, a shallow water table and a demonstrable connection with surface water. Similar considerations mean that these deposits are still heavily utilized today in the industrialized as well as in the developing world. Historically, the kanats (long underground galleries, collecting groundwater from detrital deposits along the foothills of the mountains and conveying it to the cities of the plains) of Persia and neighbouring countries are among the oldest known public waterworks (Chapter 9).

NON-AQUIFERS

Not all the rocks in the Earth's crust are aquifers. At great depths, conditions are such that open voids cannot exist, and rocks therefore have zero porosity. Before these depths are reached, conditions generally exist under which pores and fractures are so reduced in size that, although the rocks may have measurable porosity, for practical purposes they are impermeable. These rocks thus serve as the lower boundaries of the deepest aquifers.

At lesser depths, there may be rock formations that have aquifers above or beneath them but which themselves have permeabilities too low for them to be called aquifers; a confining bed is an example of such a formation. These formations frequently contain water – that is to say they are porous – but do not allow water to move through them under typical hydraulic gradients; such formations are sometimes called **aquicludes**. Other formations permit water to move through them, but at much lower rates than through the adjacent aquifers; in particular they may permit the vertical flow of water between underlying and overlying aquifers. There was a suggestion that this type of formation be called an **aquitard**. In 1972, however, the United States Geological Survey published recommendations which sought to end some of the confusion that was arising over the use (or misuse) of hydrogeological terms – an act for which the USGS deserves the gratitude of all English-speaking hydrogeologists. One of the recommendations was that terms like 'aquiclude' and 'aquitard' should be discontinued, and reference be made simply to confining beds with some description of their permeability relative to the adjacent aquifer. Whether or not this recommendation eventually wins support, terms like 'aquiclude' and 'aquitard' are common in existing literature, so it is as well to know what they mean.

These formations of low permeability are important. True, they will not yield water to wells, but they play an important part in controlling the movement of water in adjacent permeable formations. Furthermore, as we shall see in Chapter 12, there are times when the hydrogeologist or engineer deliberately seeks impermeable rocks – as suitable material in which to make a deep excavation, for example, or as a suitable location for a dam.

IGNEOUS AND METAMORPHIC ROCKS

The principal difficulty facing anyone dealing with the hydrogeology of igneous or metamorphic rocks is their extreme variability. This is largely because these rocks possess little primary porosity: in general the porosities of unweathered pieces of metamorphic or intrusive igneous rocks are less than 1%, and hydraulic conductivities of such pieces are unlikely to exceed 10^{-5} m/day. The water-bearing capacity of the rocks is therefore related almost entirely to the secondary porosity, which develops as a result of fracturing or weathering. These weathered rocks provide small but locally important supplies in many areas of the developing world.

The effects of weathering are usually limited to depths of less than 100 m, and below this fractures tend to be closed by the weight of the overlying rock. The permeability of these rocks therefore decreases with depth. There are always exceptions; mines in the Canadian Shield, for example, have recorded significant inflows more than 1000 m below the surface and significant permeability was encountered at greater depths in the Kola 'Superdeep' well (Chapter 2).

Because the extent of weathering and fracturing varies from one rock type to another, and varies with geological history and climate within the same rock type, it is almost impossible to generalize on the hydrogeological properties of these rocks. Large variations within a single rock type are possible over small areas, so that detailed investigations are necessary before any predictions can be made as to well yields or the amount of water likely to enter an excavation. Unfortunately, the cost of detailed investigations can only rarely be justified – more usually in assessing the difficulties of disposing of water from underground works than in predicting well yields for water supply. However, in recent years, the search for safe underground sites in which to dispose of radioactive waste from nuclear power stations has given great impetus to the study of igneous and metamorphic rocks (Chapter 13). Underground research facilities have been set up in places such as Stripa in Sweden, Manitoba in Canada, and in Switzerland.

Extrusive rocks tend to be even more variable than the intrusives. The dense varieties have primary porosities of less than 1%, while pumice may have porosities as high as 85%. Pumice is essentially solidified foam, consisting of lava containing numerous vesicles. Because the vesicles are rarely interconnected, the permeability of pumice is usually low.

Many hollow, generally tubular structures can form in lavas, but where high permeability exists in lavas it usually results from one of two causes: the spaces that form as a result of shrinkage into columnar structures, and the voids that occur between successive lava flows as one is deposited upon the weathered surface of its predecessor.

The weathering of lava flows leads to the formation of secondary minerals which can fill the primary pores. In many cases, therefore, the permeability of these rocks decreases with time. Some lavas (such as the basalts of the Snake River and Columbia areas of the United States) form important aquifers (see above), while others, such as the Deccan basalts of India, have low permeabilities.

SELECTED REFERENCES

Back, W., J.S. Rosenshein and P.R. Seaber 1988. *The Geology of North America, Volume 0–2: Hydrogeology*. Boulder, Colorado: The Geological Society of America.
Barrow, G. and L.J. Wills 1913. Records of London wells. *Memoirs of the Geological Survey of England and Wales*. London: HMSO.
Cohen, P.L. 1985. Ground-water development in the United States of America. *Hydrogeology in the service of man*: Memoires of the 18th Congress of the International Association of Hydrogeologists, 17–30. Cambridge: IAH.

Davis, S.N. and R.J.M. De Wiest 1966. *Hydrogeology*. New York: Wiley. (Chs 9, 10 and 11 provide a useful and straightforward account of various rock types as aquifers.)

Downing, R.A., M. Price and G.P. Jones 1993. *The hydrogeology of the Chalk of North-west Europe*. Oxford: Clarendon Press.

Driscoll, F.G. 1986. *Groundwater and wells*, 2nd edn. St Paul, Minnesota: Johnson Filtration Systems.

Helgesen, J.O., D.G. Jorgensen, D.B. Leonard and D.C. Signor 1982. Regional study of the Dakota aquifer (Darton's Dakota revisited). *Ground Water* **20** (4), 410–14.

Lohman, S.W. *et al.* 1972. *Definitions of selected ground-water terms – revisions and conceptual refinements*. US Geological Survey Water-Supply Paper 1988. Washington, DC: US Govt Printing Office.

Marsh, T.J. and P.A. Davies 1984. The decline and partial recovery of groundwater levels below London. *Proceedings of the Institution of Civil Engineers* Part 1, **74**, 263–76.

Meinzer, O.E. and H.A. Hard 1925. *The artesian water supply of the Dakota Sandstone in N. Dakota, with special reference to the Edgeley Quadrangle*. US Geological Survey Water-Supply Paper 520-E. Washington, DC: US Govt Printing Office.

Rodda, J.C., R.A. Downing and F.M. Law 1976. *Systematic hydrology*. London: Newnes-Butterworth. (Ch. 5 contains an excellent and concise account of the major British aquifers.)

Todd, D.K. 1980. *Groundwater hydrology*, 2nd edn. New York: Wiley. (See especially Ch. 6.)

<table>
<tr><td>

8

</td><td>

Springs and rivers, deserts and droughts

</td></tr>
</table>

DISCHARGE FROM AQUIFERS

Earlier chapters have shown something of how water enters an aquifer, how it is stored there and the factors that govern its movement through the aquifer. Now it is time to consider how water leaves the aquifer. One way it can leave is by abstraction from a well, and we shall see more of this in Chapter 9. But water was entering and leaving aquifers long before there were people on Earth to sink wells, so let us first look at how nature accomplishes this part of the water cycle.

Water that leaves aquifers naturally usually finds its way into river systems (although water from coastal aquifers may discharge direct to the sea). The area of land that drains to a river is called the river's **catchment** or 'catchment area'; the higher land that separates two catchments is called, in Britain a **watershed** or a **divide**. (In the United States, where the term 'watershed' is often used for a catchment area, the boundary between catchments is always called a 'divide'.) In impermeable terrain, the catchment contributes overland flow and interflow; where permeable rocks crop out within the catchment the river will usually derive some of its flow from groundwater. In the latter case, the river can be considered to have a **surface-water catchment** and a **groundwater catchment**; if the water-table relief exactly follows the ground relief, the two divides will coincide, as in Figure 8.1a. This will occur if the rocks are homogeneous and isotropic,

Figure 8.1 Recharge and discharge areas. (a) Surface-water divides (S) and groundwater divides (G) between valleys in a region composed of permeable, homogeneous and isotropic rock. (b) Flow pattern of groundwater in the catchments shown in (a). (c) Vertical components of groundwater flow imply vertical hydraulic gradients which can be measured using piezometers. Note that head decreases with depth below recharge areas (R) and increases with depth below discharge areas (D). These effects are shown by the water levels in the piezometers at R and D. In the deep piezometer at D, the head is above ground level. (Figure 8.1b after Hubbert, 1940.)

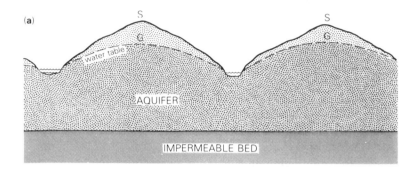

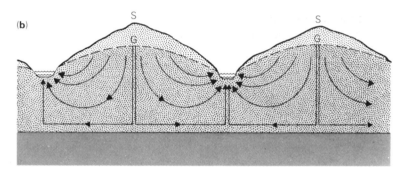

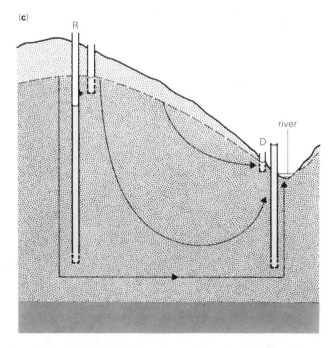

and may occur even if they are not. The groundwater flow paths for a homogeneous, isotropic catchment take the form shown in Figure 8.1b. Flow occurs from the **recharge areas**, which are on high ground, to the **discharge areas**, which are low-lying. Note that flow occurs throughout the aquifer; there is no stagnant zone at depth.

The vertical-flow components mean that there must be vertical components of hydraulic gradient. The resulting head differences between points at different levels in the aquifer can be measured using piezometers (p. 56). There is a downward component of flow (and therefore of hydraulic gradient) under recharge areas and an upward component under discharge areas (Figure 8.1c). These head differences occur even if the aquifer is perfectly homogeneous and isotropic; they may be increased by the variations of permeability with depth which are common in many aquifers. The classic 'artesian' situation (Figures 7.2 and 8.5) is an extreme case of the condition in the deep piezometer at D (Figure 8.1c).

The groundwater and surface-water divides rarely coincide as precisely as shown in Figure 8.1a. Sometimes major departures occur, as shown in the theoretical case of Figure 8.2, where the permeable bed B is effectively transferring water from the surface-water catchment area of river C to that of river D; as a result, the groundwater divide G is closer to C than is the surface-water divide S. The groundwater catchment of the River Itchen, in the Chalk of Hampshire, for example, is estimated by Ineson and Downing (1964) to be some 20% larger than the surface catchment. The most extreme examples of non-coincidence of groundwater and surface-water divides occur in cavernous limestone areas.

Water will flow from an unconfined aquifer wherever the water table intersects the ground surface (Figure 8.3). Where the flow from an aquifer is diffuse it is termed a **seepage**; where it is localized, as for example along

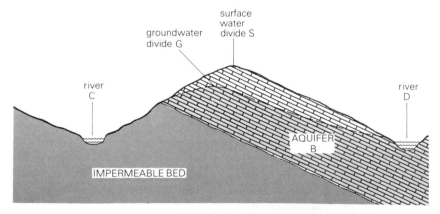

Figure 8.2 Non-coincidence of surface-water and groundwater divides. In this theoretical example, transfer of groundwater, through bed B from the catchment of River C to the catchment of River D leads to the non-coincidence of the surface and groundwater divides.

(a)

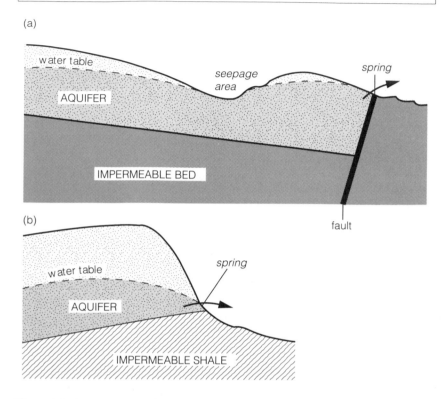

(b)

Figure 8.3 Common occurrences of springs and seepage areas.

a fault or fissure (Figure 8.3a) it is called a **spring**. It is common to find lines of springs or seepages where permeable sandstones or limestones form high ground and rest on less permeable rocks such as shales or clays (Figure 8.3b); these spring lines are often used by field geologists as a guide when mapping the boundary between two such formations.

The largest known spring in the world issues from a limestone at Ras-el-Ain in northern Syria. Its flow, at a rate of about 40 m^3/s, helps to sustain the flow of the Euphrates via its tributary, the Khabour.

Water will flow from a confined aquifer where the potentiometric surface is above ground level *and* where there is locally some form of permeable path through the overlying confining bed (Figure 8.4). However, it is much more common for groundwater to leave a confined aquifer by percolating slowly through a confining bed into permeable material (Figure 8.5). This is one of the mechanisms ignored by the classical explanation of artesian aquifers (pp. 70–1).

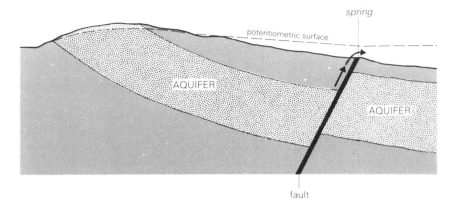

Figure 8.4 Discharge from a confined aquifer. A fault plane can provide a permeable path along which groundwater can discharge from a confined aquifer.

WHY RIVERS KEEP FLOWING

As Figure 8.1b shows, discharge of groundwater occurs where the bottom of a river valley lies below the water table. The discharge may take place through the bed or bank of the stream or river and so may not be visible, but such discharges account for the greatest proportion of flow from aquifers. A river that receives water from an aquifer, like that in Figure 8.1b, is termed a **gaining stream**. For a river to flow throughout the year, even during long periods without rainfall, it must have a source of water other than surface runoff or interflow. This water, which sustains the river throughout dry weather, is present, though less apparent, at other times; it is termed **baseflow**. Baseflow can be provided by groundwater discharge from an aquifer, from surface-water storage (as in the case of a river that flows from or through a lake) or from the melting of glacier ice or of snow

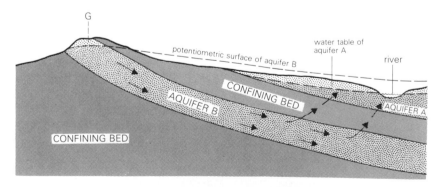

Figure 8.5 Discharge from a confined aquifer. Natural discharge from a confined aquifer (B) across a confining bed. If there were no discharge from B, its potentiometric surface would be horizontal from G eastwards.

which is present throughout most of the year. The first of these sources is the most common, and many writers use the terms 'baseflow' and 'ground-water discharge' as though they were synonymous.

It may be that a stream flows across permeable material but that the bed of the stream is higher than the water table. In such a case, unless the stream bed is itself impermeable (perhaps floored with clay, for example) water will flow from the stream to the aquifer; in this case, the stream is called a **losing stream**. It is possible for a river to be a gaining stream over one part of its length and a losing stream over another part; or for the same stretch to be a gaining stream at some times and a losing stream at others, as the water table rises and falls.

A river that flows throughout the year, every year, is called a **perennial stream**, and a river that flows only occasionally, perhaps for hours or days in several years, is called an **ephemeral stream**. Ephemeral streams usually occur in semi-desert regions, where rainfall, though unpredictable and unusual, may be heavy and localized in the form of storms. A sudden downpour produces sufficient surface flow or interflow to sustain the stream for a short time, before its flow is lost by evaporation or infiltration. Ephemeral streams rarely have well-defined channels and are never gaining streams.

Finally there are **intermittent streams**. These flow for part of each year, usually during or after the season of most rainfall, or as a result of snowmelt. One occurrence is where the water table, after infiltration in the winter, rises above the bed of the upper reach of a river (Figure 8.6); as the height of the water table declines, the source of the river travels down from A to B. Downstream of B, which is termed the **perennial head**, the stream is perennial. The situation shown in Figure 8.6 is common in the Chalk downlands of southern England where, because of the low specific yield of the Chalk, large fluctuations occur between the spring and autumn positions of the water table. These intermittent portions of Chalk streams are called **winterbournes**, which explains the common occurrence of 'winterbourne' or simply 'bourne' in the place names of the area.

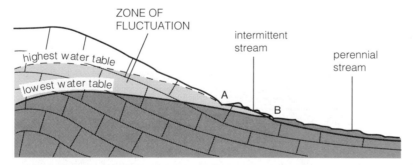

Figure 8.6 An intermittent stream. An intermittent stream may occur in response to a changing water table.

A common statement is that groundwater supplies about one-third of all
the water used for public supplies in England and Wales; by 'groundwater'
the authors mean water pumped from wells or collected from springs. The
statement is perfectly true – it is made in this book – yet it underestimates
the importance of groundwater in supplying the needs of England and
Wales, because it ignores the contribution that groundwater makes to the
flows of our major rivers, many of which are used as sources of supply. As
we have just seen, rivers can only flow throughout the year, even in a
temperate climate like that of Britain, if they have a source of baseflow. In
Britain, the baseflow component of all our major rivers is derived from
groundwater. It therefore follows that, during dry periods, the water
abstracted for public supply from rivers is indirectly derived from aquifers.
To estimate the importance and quantity of this baseflow it is necessary to
know something of how river flows are measured and analysed.

MEASURING RIVER FLOWS

The **discharge** of a river or stream is the volume of water flowing past a given
point in a unit of time; it is therefore the cross-sectional area of the flow
section multiplied by the speed with which the water is flowing. If the shape
of the river channel is known, then the cross-sectional area can be determined at any time provided that the depth of water in the river is known at
that time (Figure 8.7). The depth of water is usually measured in terms of
the height of the water surface above a reference or datum level; this height
is called the river **stage**, and a graph of river stage against time is called a
stage hydrograph. Stage is most simply measured using a vertical graduated
post (a **staff gauge**) set in the river bed or against the bank; these staff gauges

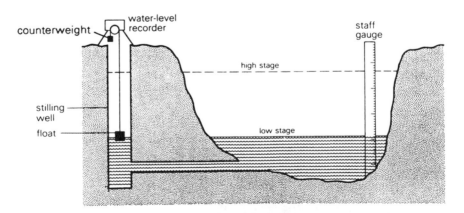

Figure 8.7 Cross-section of a river channel at a gauging station. The stilling well
eliminates ripples and wind effects, to provide a smooth surface for water-level
measurement.

can be seen at intervals along most British rivers. More usefully, recorders can be installed which measure water level continuously and which produce a hydrograph automatically (Figure 8.7). The datum from which the stage is measured is not necessarily the bed of the river at that point; often stage measurements at several points along a river are made relative to a common datum.

Stage measurements are of importance to engineers concerned with river management, for purposes such as navigation and for flood prediction and control. Those – including hydrogeologists – whose primary concern is water resources are more interested in the river discharge, for which it is necessary to know the speed of flow as well as the stage.

Flow speed can be measured in a variety of ways but it is of little use to know the speed and the stage – and therefore the discharge – at only one time. Much more valuable is the record of flow plotted against time that is depicted in a **discharge hydrograph**. As we have seen, it is relatively easy to measure the river stage and to produce a stage hydrograph. If the hydrologist or engineer can produce a graph or formula that relates discharge to stage for the entire range of flows which may occur at that point on the river, then subsequently he or she can derive the discharge value whenever it is wanted simply by measuring river stage. The point on the river where this is done is called a **gauging station**, and Figure 8.8 shows the form of a **stage–discharge** relationship or **rating curve** for such a station.

There are two principal ways of establishing the stage–discharge relationship. The first is by choosing a suitable stream section, and then measuring the cross-sectional area and speed at various stream stages and so constructing a rating curve like that in Figure 8.8. A gauging station that uses this method is a **velocity–area** station. A simple (though not very

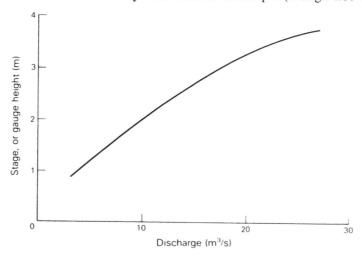

Figure 8.8 A stage–discharge graph. The form of a stage–discharge graph or rating curve for a river gauging station.

accurate) way of measuring flow speed is to drop floats into the water and time them over a known distance. A more accurate and common method is to use a **current meter**. This is mounted on a rod or on a weighted cable so that it can be lowered into the river flow to any desired depth. The rod or cable can simultaneously be used to measure the depth of the stream.

Traditional current meters use a rotating propeller or an anemometer-like vane to measure the speed of the water flowing past them, but the latest instruments rely on electromagnetic principles. Natural water conducts electricity. The water flowing past the meter therefore represents a moving electrical conductor. When an electrical conductor moves through a magnetic field, an electrical potential difference (a 'voltage') is produced. Magnetic coils in the meter generate a magnetic field, and a potential difference is produced in the moving water. This potential difference is detected by electrodes in the meter and is dependent on the speed of flow, which can therefore be evaluated. With no moving parts to create friction or be damaged, electromagnetic meters are both accurate and robust.

By making measurements of speed and depth across the river section, the rate of flow can be calculated for that water level or stage. In practice, measurements may have to be made at several depths at each point as the speed usually varies with depth as well as with distance from the bank. On very large rivers measurements may be made from boats or by lowering the meter from a specially constructed cableway. And all the measurements must be repeated, a great many times, so that the full rating curve can be produced.

At some modern gauging stations, current meters are not used. Instead ultrasonic signals are sent diagonally across the river, underwater, between special transmitters and receivers. Going in one direction the signals are speeded up by the movement of the water, and in the other direction they are slowed; the difference between the two sets of travel times enables the flow speed to be calculated.

The problem with all of these velocity–area stations is that the relationship between stage and discharge changes with time, as a result of river bed or bank erosion, growth of aquatic plants and other factors. The second way of establishing the stage–discharge relationship avoids many of these problems by building an artificial structure – a weir or a flume – across the river. A **weir** is essentially a wall over which the water flows, while a **flume** is essentially a throat – a reduction in width or depth, or both – through which the water flows with increased speed. For both weirs and flumes, provided that the structure is built in a standard way and that its dimensions are known, the flow through or over it can be calculated from formulae if the height of water is known. As the height is known from the river stage measurement, the latter can be converted directly to a discharge value.

For reasons of cost these structures cannot be built on large rivers, for which velocity–area stations remain the only practical means of measuring discharge. Flumes tend to be more expensive than weirs; they are favoured on small upland streams, particularly those that carry a lot of silt or debris. This material can build up behind a weir and change its characteristics, whereas the increased speed of the water as it passes through the reduced cross-section of the flume tends to keep the structure clear.

It should be noted that many weirs are built primarily for purposes other than flow measurement. On larger rivers, these purposes include prevention of flooding by controlling the rate at which water is allowed to flow from one stretch of the river to the stretch downstream, and the maintenance of sufficient depth of water for navigation.

HYDROGRAPH ANALYSIS

Having gone to such lengths to collect this streamflow information, how can we use it? Figure 8.9 shows a theoretical stream-discharge hydrograph for a river draining a catchment that is underlain by some permeable and some relatively impermeable material. A few years ago, a classic exercise for hydrologists or hydrogeologists would have been to *analyse* such a hydrograph, a process which involved deciding how much of the flow at various times was derived from groundwater flow, how much from surface runoff and how much from interflow. The theories behind this analysis were based largely on the work of R.E. Horton in the USA, who, in the 1930s, put forward the idea of infiltration capacity being a major control on the way rainfall is disposed of over a river catchment. With the realization that overland flow and surface runoff were rare events in many catchments, objections to Horton's description of streamflow began to be raised. New theories were put forward in the 1960s, principally by J.D. Hewlett, also in the USA. As outlined in Chapter 3, these theories place greater emphasis on interflow, and modern hydrograph analysis tends to recognize two components – a **baseflow** component, consisting usually of groundwater flow and slow interflow (plus, in some cases, meltwater and water from surface storage), and a **quickflow** component, derived from rapid interflow, any surface runoff and any rain that falls directly on the river channels. The division tends, therefore, to be based on the length of time the various components take to reach the main drainage channels, rather than on their route. As will be seen shortly, this is not always satisfactory to the hydrogeologist.

The hydrograph segment in Figure 8.9 commences (time A) at the end of a long period without rain, when we can assume that all the flow in the stream was baseflow, derived largely from groundwater storage. As this groundwater discharge took place, the amount of groundwater stored in the aquifers decreased, leading to a fall in the potentiometric surfaces and a reduction in the hydraulic gradient and hence in the flow of ground-water to the river; as a result the hydrograph (GH) shows a gradual decrease in flow.

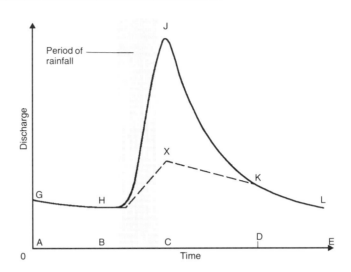

Figure 8.9 Hydrograph analysis.

Between B and C a period of rainfall occurred. Some of the rainfall was disposed of as quickflow, reaching the surface watercourses rapidly and leading to an increase in river discharge to a peak value at J. This increased flow then began to diminish, and it can be assumed that by time D the flow was once again all baseflow. The discharge is now greater than it was before the rainfall, however, indicating that there has been an increase in the amount of water stored in the aquifers; some of the rainfall has therefore infiltrated into the aquifers.

The suggested division of the total discharge into baseflow and quickflow is shown by the dashed line in Figure 8.9. This dividing line is termed the **baseflow-separation curve**; it could be drawn in a variety of ways and according to a variety of rules, the final choice invariably being subjective. The possibility exists nowadays of drawing it by computer; computer methods achieve consistency, but they still have subjectivity built into the programs they use, because the programs have to be written by people.

The discharge hydrograph is a graph of discharge (volume of water per unit of time) plotted against time. It therefore follows that the area under the graph between any two times (e.g. the area AGHJKLE in Figure 8.9) represents the total volume flowing past the gauging station during the time interval AE. Similarly, the area beneath the baseflow-separation curve (the area AGHXKLE, for example) represents the volume of baseflow passing the gauge in that time interval.

Hydrograph analysis thus provides a way of knowing how much ground-water is flowing from a catchment area upstream of the gauging station in a given time period – a year, for example. This is interesting, but much more

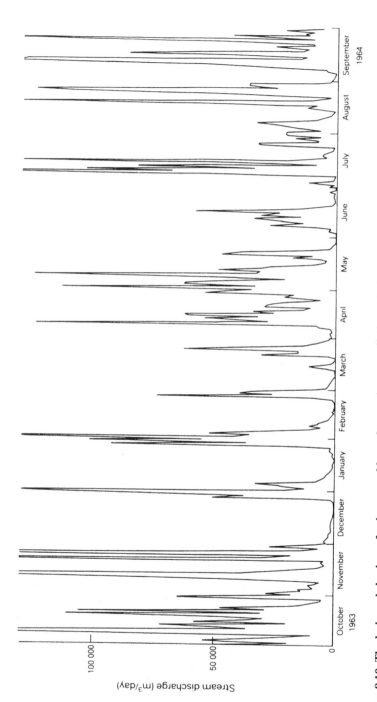

Figure 8.10 The drainage behaviour of an impermeable catchment. A stream-discharge hydrograph for a stream draining a catchment area underlain by metamorphic rocks in the Western Highlands of Scotland. (Peak discharges are not shown above 130 000 m^3/day.) (Reproduced by permission of the Director, British Geological Survey, and based on data supplied by North-West Scotland Hydro-Electric Board.)

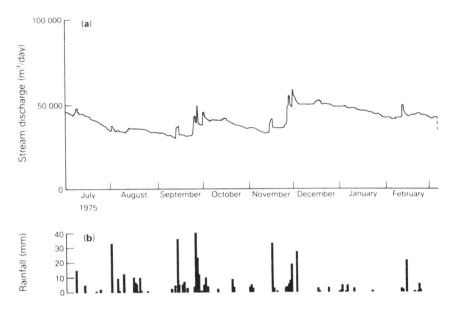

Figure 8.11 The drainage behaviour of a permeable catchment. (a) A stream-discharge hydrograph for a stream draining a catchment area in Hampshire underlain by Chalk. (b) Rainfall at a nearby rain gauge (amounts less than 1 mm have been omitted). (Based on data supplied by Southern Water Authority.)

significant is the fact that if groundwater is leaving the catchment it must have entered it – in other words, the volume of water leaving the catchment as baseflow must have entered as infiltration. Dividing this volume by the catchment area enables us to express the infiltration in millimetres, so that we can compare it with rainfall and evapotranspiration figures. Doing this calculation over one year can be misleading, as there may be changes in the amount of groundwater in storage in the aquifers, especially if there have been unusual rainfall conditions. It is therefore advisable to average the results over a period of at least five years and preferably more than ten. Used sensibly, this is the best general technique for assessing the infiltration over an area. However, it can work satisfactorily for the hydrogeologist only if the separated component represents infiltration that reaches the aquifers; this is why an arbitrary division into baseflow and quickflow may not be entirely satisfactory. It is also essential to know the area of the catchment contributing the baseflow; if the groundwater and surface-water catchments do not coincide this may be difficult to determine.

Figure 8.9 was a hydrograph for an imaginary catchment containing permeable and impermeable material. Figure 8.10 shows a real hydrograph for a Scottish stream draining a catchment underlain almost entirely by impermeable metamorphic rocks (mica-schists). The inability of rainfall to infiltrate to any depth, and the lack of groundwater storage, means that a

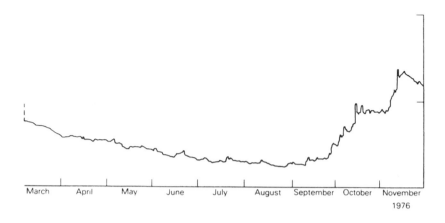

rainfall event produces a rapid rise in streamflow, which is followed by an almost equally rapid decline at the end of the rainfall. In dry weather the flow almost ceases, being sustained largely by minor baseflow contributions from peat and superficial deposits. This type of river response is termed 'flashy'; in areas like this, flooding is likely after heavy rain, and catchments are characterized by numerous streams and tributaries.

In contrast, Figure 8.11 shows a hydrograph from a river draining a catchment underlain by Chalk in southern England. Here there is hardly any surface runoff or interflow, because almost all the rainfall infiltrates into the Chalk and emerges as baseflow.

In permeable catchments like this one there is little in the way of surface drainage; virtually all streams are gaining streams whose valleys intersect the water table. Many factors other than geology influence the response of streams to rainfall, but aquifers exert a stabilizing influence on streamflow, taking water into storage during periods of heavy rainfall (thereby reducing the possibility of flooding) and releasing it slowly during dry weather, thus maintaining streamflow which is available for water supply, dilution of sewage effluent, navigation, recreation (including fishing), and so on. Statistics that consider only the amounts pumped directly from aquifers can therefore seriously underestimate the importance of groundwater.

WHAT HAPPENS IN A DROUGHT?

The importance of groundwater in maintaining streamflow becomes more apparent in a drought. Having said that, it must be admitted that there is no agreed definition of the word 'drought', because dry weather affects different people in different ways.

In Britain the Meteorological Office defines a drought as a period of at least 15 consecutive days without measurable rainfall. For this purpose 'measurable' means in practice more than 1/100th of an inch or, these days, 0.25 mm. The way that such a period affects people will depend very much on who we are and when it occurs. In general people are much more likely to notice it in summer, when grass may begin to turn brown and gardens need watering, than in winter. A dry period in spring may please most people but could be disastrous to farmers, who want plenty of soil moisture for their growing crops. Similarly the farmers may welcome a relatively dry winter, but this is precisely the time that water engineers want rain.

Contrary to popular opinion, 'drought' in summer makes little difference to our water supplies, except by increasing demand for garden watering and spray irrigation – demands that, although conspicuous, represent a relatively small proportion of our annual national water consumption. This is because the high soil-moisture deficits that develop in spring and summer mean that most rainfall is taken up by the soil anyway, with little left to flow into streams and reservoirs or recharge aquifers. Summer droughts are only really significant in northern and western Britain. In these areas – usually – summer rainfall is great enough, and temperatures are low enough, for there to be some surplus of rainfall over evapotranspiration for much of the year; in consequence, a major reduction in summer rainfall may be important.

In southern Britain, in the months from April to September, average potential evapotranspiration is greater than average rainfall. This means that a soil-moisture deficit develops usually by the end of April or early May. Thereafter, until September or October when lower temperatures and shorter days bring an end to most plant growth and evaporation, only exceptionally will the rainfall be sufficient to wipe out the soil-moisture deficit and allow infiltration. It is in autumn, usually, that the soil-moisture deficit is eradicated, and in winter and early spring that falling rain can escape the clutches of the soil to form recharge to aquifers or runoff that can replenish reservoirs. If that rain does not fall in autumn, the soil-moisture deficits can persist into winter. Then recharge is greatly reduced.

It is for these reasons that water engineers in Britain worry far more about dry periods in winter than in summer, and why, even though there may not have been a strict meteorological drought in winter, water resources the following year may be under threat. This also explains why meteorologists, agriculturalists and hydrologists all have their own conceptions of drought. The general public tend to welcome dry weather until it persists to the point where it interferes with their lives – perhaps by leading to a shortage of fresh vegetables, or to a ban on watering of gardens, or (in Britain very exceptionally) to a rationing of water supplies to domestic consumers.

THE DROUGHT OF 1975–76

In 1975 and 1976 there occurred over most of Britain and large parts of Western Europe a pattern of weather that was accepted by all those who experienced it as a drought, and in parts of the affected area the most severe drought since records began. This consensus of opinion makes this drought, although not typical, a useful one to consider as an example.

The 1975–76 drought in Britain was not a single unusual event, but a combination of unusual events. The winter of 1974–75 was somewhat wetter than average, but from May to August 1975, England and Wales received only about two-thirds of the average rainfall for that period. (The 'averages' here refer to average precipitation during the period 1916–50, which is a standard period for this purpose.) Relatively dry weather continued throughout the winter of 1975–76 – one of the driest winters of the last century – with England and Wales receiving little more than 60% of average precipitation. The hot dry summer of 1975 resulted in high rates of potential evapotranspiration, leading to the development of large soil-moisture deficits over most of Britain – deficits that were slow to diminish during the following dry winter.

By February 1976, soils in many areas had only just about returned to field capacity, meaning that little or no extra water was available as recharge to aquifers. This meant that in many areas groundwater storage and therefore groundwater levels (potentiometric surfaces) continued to decline during the winter of 1975–76, instead of showing the usual rise. Fortunately, the wet weather of the winter of 1974–75 had resulted in high groundwater levels in many aquifers at the beginning of 1975, so that despite the dry summer of 1975 and the lack of recharge during the succeeding winter, groundwater storage was not seriously depleted.

Spring and summer of 1976 continued to be dry (Figure 8.11). The summer was also exceptionally hot and sunny, with record high temperatures during late June and early July and record sunshine levels for August at many observation stations. These conditions favoured high evapotranspiration, and soil-moisture deficits soon reached the point at which there was little or no moisture available for plant growth.

Groundwater levels continued to decline, so that springs and seepages either ceased to flow or flowed at a diminished rate; since all other flow had disappeared, this reduction in baseflow resulted in many streams and rivers almost ceasing to flow. Similarly, the lowering of water tables meant that many shallow wells became dry, in some cases for the first time in living memory. In general, however, deep wells such as those used for public supplies proved reliable. In contrast, some areas of the country dependent on rivers or surface-water reservoirs for their water supplies suffered shortages, with many reservoirs suffering from a combination of reduced recharge in the 1975–76 winter and high evaporative losses in the dry summer of 1976. In some of these areas water rationing was imposed.

One possible effect of droughts is that the decline of groundwater storage in unconfined aquifers can sometimes result in a water table being lowered below the bed of a river, so that what is normally a gaining stream becomes a losing stream. An apparent example of this occurred in the drought of 1975–76, resulting in the much-publicized Thames 'leak'. During the driest period of 1976, it was discovered that the flow of the Thames between Eynsham (near Oxford) and Dorchester, some 40 km downstream, was decreasing. A study indicated that much of the decrease could be accounted for by evaporation, but that the remainder appeared to be caused by water seeping from the river into the adjacent aquifer.

Whatever else the drought did, it drew attention to the fact that water is an important commodity which, even in a country like Britain with a temperate and humid climate, cannot always be taken for granted. With the general public being made uncomfortably aware of this fact, it was not long before politicians at all levels and of all political shades were competing with each other for news coverage as they either praised or criticized the water authorities for coping or not coping with the problem. The news media were able to fill in the 'silly season', reporting all these aspects.

But all things come to an end, and the drought ended with a September and October that were the wettest in England and Wales for over 250 years, and with a generally wet winter. Soil-moisture deficits disappeared, aquifers and reservoirs were replenished, and most people went back to washing their cars and not worrying about water shortages – until the next time.

THE DROUGHT OF 1984

For much of Great Britain, 'the next time' came in 1984, though it was a drought of a different kind from that of 1975–76. The wet winter that brought an end to the 1975–76 drought was followed by a succession of winters with rainfall generally above average. Initially, the winter of 1983–84 looked set to continue that trend; the summer of 1983 was hot and dry, but September–December 1983 was a generally wet period although rainfall in southern and eastern England was below average. In January 1984 the United Kingdom, taken as a whole, experienced more than one-and-a-half times the normal rainfall. By the end of that month all major reservoirs were at or very close to capacity, and groundwater levels were generally above the average for that time of year, in some cases by a substantial margin.

Over the next few months the pattern changed, with a warm, dry spring followed by a hot, dry summer. Unlike 1975–76, when the impact of the drought was felt strongly in the normally dry south and east of Britain, the rainfall deficiency in 1984 affected all of the United Kingdom, with the greatest deficits in Scotland, Northern Ireland, and in the upland areas of northern England, the south-west and Wales. All of these areas are heavily dependent on surface reservoirs for water supply; by August many of those reservoirs were looking ominously short of water. In consequence, people in all walks of life in places that normally take water for granted were faced with a new experience.

In August a depression brought widespread rainfall to England and Wales, but this was generally taken up by the high soil-moisture deficits. By the end of the month Thirlmere and Haweswater in the English Lake District were reduced respectively to 13% and 16% of their capacities. In much of north-west England, reservoir stocks were sufficient for only about another 50 days' supply.

In the lowlands, much more dependent on groundwater for their supplies, most residents were probably unaware of the existence of a drought, for three main reasons. First, the rainfall deficit (the amount by which the rainfall fell short of the long-term average) was much less in the lowland areas than in the uplands, *even though* the uplands still received *more rain* than the lowlands. Second, the drought was predominantly a spring and summer drought; as little recharge to British aquifers occurs after March anyway, the shortfall was not significant. Third, the volume of water stored in aquifers is very much greater than the amount held in reservoirs. A fall of just 1 m of the water table in the unconfined Chalk aquifer, for example, will release more than 200 million m^3 of water; this is equivalent to the storage capacity of Kielder Water, the largest water-storage reservoir in Britain.

The 1984 drought was effectively broken for most of Britain in September, with rainfall for that month approaching the total for the previous five months in some areas. The exceptions were Northern Ireland and parts of southern Scotland, which received sporadic but generally low rainfall in August and, in some areas, little in September. Some Scottish reservoirs were at only 20% capacity by mid-September; Lough Neagh in Northern Ireland, whose large but shallow expanse was particularly affected by evaporation, was also very low. However, heavy and prolonged rainfall in October and November put an end to most problems, with reservoirs generally at or near capacity by the end of the year.

The droughts of 1975–76 and 1984 demonstrated the vulnerability of surface-water reservoir storage to sustained dry periods, and the fact that groundwater sources generally prove more reliable in these circumstances. It might be thought that two such sharp lessons in the space of less than ten years would be remembered for a long time but, as the next drought showed, it is remarkable how fickle the human memory can be.

THE DROUGHT OF 1988–92

The decade of the 1980s was the wettest recorded in Britain. It was also a decade of extremes. There is normally a marked difference between the rainfall in the north and west of Britain, and that in the south and east. The hard rocks that underlie most of the north and west (Figure 7.6) generally form high ground, upon which the moist warm air that sweeps in from the Atlantic – from which the prevailing winds come – deposits much of its moisture. The generally younger, softer rocks of the south and east form lower ground that lies in a 'rain shadow'. Thus the north and west on average receive most rain and snowfall (in excess of 2500 mm a year over large areas) and the south and east receive the least – less than 500 mm a year in a few places. On top of this, the cool cloudy north and west lose little of that water to evapotranspiration, whereas in large parts of the sunnier south, in a typical year, as little as 150 mm of **effective rainfall** is left to infiltrate aquifers or provide streamflow.

At the turn of the decade, these normal differences became greatly exaggerated. The 1988–92 drought showed some interesting parallels with that of 1975–76. Just as the winter of 1974–75 had been a good one for recharge, so too the spring of 1988 found groundwater levels higher than normal in many parts of England – in some cases setting new records (Figure 8.12).

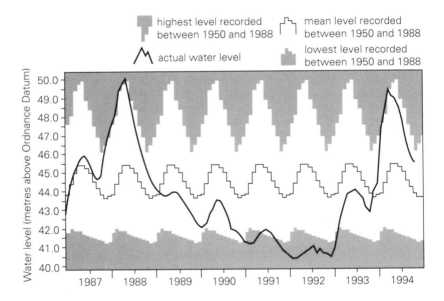

Figure 8.12 Water-level fluctuations in a well in the Chalk aquifer in Norfolk. Measuring datum is 80 m above Ordnance Datum (OD); depth 46 m; the Upper Chalk is overlain by clays and sands. The solid line is the record of the way the water level changed in the Chalk in this area before, during and after the drought of 1988–92. The other lines show the monthly maximum, minimum and mean water levels that had been recorded in the well between 1950 and 1988. The water level is an indication of the amount of water stored in the aquifer. Fortunately the drought began when the storage was at an unusually high level, and was followed by a return to high levels. The general form of the record is typical of many wells in the Chalk in eastern and southern England during this period. (Reproduced by permission of the Director, British Geological Survey.)

The origins of the drought, like those of its predecessor in 1975–76, lay in a change in the normal atmospheric circulation patterns over Western Europe. The depressions that normally bring much of the winter rain to the British Isles and Western Europe were shifted to a more northerly route by stable and nearly stationary high-pressure systems. In consequence most of Scotland experienced a marked increase in rainfall while much of southern Britain faced a period of unusually dry winters combined with hot dry summers, that of 1989 being especially notable. The first shortfall came in the autumn and early winter of 1988–89, but it was followed by a wet spring. Recharge to aquifers was reduced and also delayed. The combination meant that groundwater levels in many areas, although low for much of the winter, were back to near normal by May 1989. The low winter levels however meant that streamflows – especially of winterbournes (p. 103) were much depleted, the start of a worrying trend.

Initially the drought of the late 1980s was very like that of 1975–76 in its effects. South-west England was again badly affected, so that by August 1989 there was renewed talk of water rationing. Areas dependent on groundwater were again relatively well provided for, except for a strip down the east coast of England and part of eastern Scotland, which had not received the high spring rainfall. Autumn 1989 brought welcome rain to much of the west, alleviating the problems of south-west England, but not to the east, where once again soil-moisture deficits were persisting.

The drought was then punctuated in most areas by a remarkably wet period in December 1989 to February 1990, when much of the country experienced its wettest three such months since 1915. From a hydro-geological viewpoint the rainfall intensity was too great – a great deal of surface runoff and interflow was generated, depriving aquifers of recharge – and the period rather too short to be ideal. Even so, the rain was very welcome to water engineers, though some eastern areas continued to receive less than average. Thereafter a dry spring set a pattern for much of the next two years, during which groundwater levels in many parts of eastern England fell to their lowest recorded levels (Figure 8.12) and many streams and springs ceased to flow. Discharges on some eastern rivers were below average for almost the whole of the four years.

Another similarity of the 1988–92 drought with the drought of 1975–76 was that both roughly coincided with periods of major reorganization of the water industry in England and Wales. In 1975 the industry was adjusting to the reorganization into regional water authorities that followed the Water Act of 1973. In 1988–92 it was experiencing the major upheaval that came with the Water Act of 1989, especially the privatization of the water-supply and sewage-disposal functions of those water authorities and the setting up of the National Rivers Authority. Perhaps this explains some of the illogical thinking that was in evidence for much of the time. This took the form: south-east England is short of water; south-east England relies on groundwater; therefore south-east England is short of water because it relies on groundwater.

It is understandable that the general public found it hard to understand the severity of the drought – it is difficult to believe that a downpour that can soak people to the skin is insufficient to overcome a soil-moisture deficit and contribute to recharge – but it is less easy to excuse the way the lessons of 1976 and 1984 were forgotten by politicians and some senior people in the water industry. A recurrent theme was the claim that over-abstraction of groundwater was damaging streams and wetlands. The National Rivers Authority identified some catchments where there was excessive pumping of groundwater for public supply, leading to a reduction in streamflow that was made more apparent by the drought. Unfortunately this led many conservation groups to suppose that any stream that dried up had done so not because we were living through a most unusual drought but because a water company was abstracting too much groundwater (Chapter 12).

Political interest and media coverage were if anything more intense and often no better informed than in 1976. It was interesting to see the way that public opinion changed in response to these various factors. In the 1970s attempts by water authorities to build reservoirs were usually met with intense opposition from local people and from environmental and farming interest groups; all were usually united in opposition to what they saw as destruction of the countryside – often involving prime agricultural land or areas used for leisure. Groundwater schemes were promoted as being far less intrusive and having no harmful side effects.

By 1990 people in Europe were increasingly concerned about over-production of food; with talk of farmland having to be taken out of production, it no longer seemed to matter if some of it was covered by reservoirs. Growing affluence for at least some people led to a rise in the number of people taking up water sports and looking for large stretches of water near to home to indulge in them. Water suppliers cashed in on this trend by making reservoirs much more accessible to the public. Meanwhile the drying up of streams, allegedly as a result of over-abstraction of groundwater, caused increased opposition to groundwater abstraction, with trout fishermen on the Chalk streams being an especially wealthy and therefore influential minority. Suddenly, groundwater schemes were bad and reservoirs were good.

Where the 1975–76 drought ended spectacularly in the autumn of 1976, the 1988–92 drought went out with something of a whimper. By spring 1992 groundwater levels in many areas of eastern England stood at the lowest on record, and rainfall totals for the preceding four-year period were also at or close to record minima; compared with the long-term average, the deficiency of rainfall in much of East Anglia over that four years was the equivalent of almost a full year's rainfall. But spring 1992 was wet, and so were the summer and early autumn. For the first time since 1987, winter in these areas began with the soil-moisture deficit reduced to a level where reasonable recharge could be expected. By early 1993 it was possible to say that the drought, if not actually over, was certainly reduced in severity.

The following autumn and winter saw exceptional rainfall for much of England, and correspondingly large amounts of recharge to aquifers. In early 1994 groundwater levels in much of the Chalk were the highest recorded, and there was serious flooding in Chichester as the high water table in the Chalk led to extreme flows in the River Lavant.

The effects of the drought, like that of 1975–76, were felt beyond southern Britain. Serious effects were felt in France, Germany, Italy, Spain and Greece, among others, and in some of these countries the drought is not over at the time of writing.

LIVING IN A GREENHOUSE

The 1980s and early 1990s have seen some extremes of weather. Droughts have been or are being experienced in Europe, large parts of the United States, southern Africa and Australia. In many cases the droughts have been punctuated or terminated by catastrophic floods. There has been an increase in the frequency and severity of hurricanes and similar storms, such as those that affected southern England in October 1987 and January 1990. These events have led to suggestions that the climate of the Earth is changing. The rise in the average global surface air temperature by between 0.3 and 0.6 °C over the last hundred years is seen as evidence for such a change.

A great deal of discussion has centred around something called the **greenhouse effect**, and the suggestion that human activities have released gases – especially carbon dioxide – into the atmosphere that have triggered this effect. The result is something called **global warming**. Several points need to be understood about the greenhouse effect. To begin with, it is not new. It has been around almost as long as the Earth itself.

The greenhouse effect works because the Sun radiates most energy as visible light – which is doubtless why we have evolved eyes that are sensitive to that part of the spectrum – or at roughly similar wavelengths. Gases in the Earth's atmosphere allow energy at these wavelengths to pass through and warm the Earth's surface, but hamper the longer-wavelength energy radiated by the much cooler Earth from escaping into space. The net effect of this is to warm the surface and atmosphere of the Earth to the point where life is possible.

Second, carbon dioxide is not the most important of the gases – the so-called greenhouse gases – responsible for the effect. Water vapour is by far the most important greenhouse gas in the Earth's atmosphere. Carbon dioxide has claimed attention because the burning of fossil fuels has increased its concentration in the atmosphere to the point where it would be expected to have an effect on climate. Also, we know that the concentration of carbon dioxide in the atmosphere has varied in the past, and that higher concentrations have generally coincided with higher temperatures. We know how the concentration has varied by studying bubbles of ancient air trapped in the ice-caps of Greenland and Antarctica. Other greenhouse gases released as a result of human activities are methane, oxides of nitrogen and chlorofluorocarbons (CFCs). The introduction of these gases into the atmosphere, or the increase in their concentration, can be expected to cause an increase or enhancement of the greenhouse effect. This **enhanced greenhouse effect** or EGE is what should really concern us.

Climate, simply defined, is the long-term average weather. Meteorologists use computers to predict the weather of a region for a few days into the future, with varying success because there are so many variables that they need to consider. In the same way, climatologists can use computers to predict how the climate – the long-term weather – will vary in the future in response to changes in variables such as the concentrations of greenhouse gases. They have a difficult task because they have to model the whole of the Earth's atmosphere, and they have to look much further into the future than is normally done with weather forecasting. To keep their computer programs to a reasonable scale, they must divide the atmosphere into fairly large blocks – so large, for example, that just four or six will typically cover the British Isles. In consequence, the predictions are at present rather crude.

What the climatologists do predict, however, is that the increase in greenhouse gases should make the Earth warm up, just as it seems to be doing. So does this confirm that global warming as a result of the EGE is a reality? Unfortunately, life is never that simple. A major complication is that the Earth's climate is not naturally constant. The weather varies from day to day, year to year and – as far as we can tell from our limited records – from century to century. Although we can see a change, and that change is very much what we would expect from the EGE, we cannot be sure that it is not just part of this natural variability. Conversely, of course, we cannot be sure that the EGE is not warming the Earth even more than we predict, with the effect being partly offset at present by natural cooling as part of this variability. What we can say is that the changes we can see happening in the Earth's climate are in line with the predictions of global warming based on the EGE; they do not necessarily confirm those predictions.

Notwithstanding the recent problems, the general trend is for the British climate to get wetter. There is also a general tendency for rainfall to increase in winter and decrease in summer. These statistical trends are also in line with the predictions of global warming, and from the water-resources viewpoint look a good thing. But they do mean that the country will be increasingly vulnerable to the rogue years that defy the statistics, and leave us with an exceptionally dry winter in between the drier summers.

PERENNIAL DROUGHT – DESERTS

There are some areas of the world where drought is not an unusual occurrence, as it is in Western Europe, but is the normal state of affairs; these areas are the **deserts**. As in the case of droughts, there is no generally accepted definition of a desert. Regions with low precipitation (less than 250 mm per year) can be found at most latitudes, but the areas that most people would regard as deserts have low precipitation and lie between latitudes 15° and 50° north and south of the Equator.

These areas are characterized by their dryness, lack of vegetation and sparse population. Some are sandy, with or without dunes; some are covered with gravel, and others are predominantly bare rock. Surface drainage in true deserts is virtually non-existent, being present only after heavy rainstorms which may occur less than once in a hundred years; in semi-deserts, such as some of the intermontane regions of the United States, rainfall is more frequent and gives rise to ephemeral streams. Semi-arid areas occur in North Africa, on the edge of the Sahara, and here these ephemeral watercourses are termed **wadis**. Some recharge to aquifers beneath deserts may take place through the beds of temporary watercourses; elsewhere, the rainfall will usually be absorbed to satisfy the soil-moisture deficit before it can infiltrate far into the ground.

Many of the world's large arid and semi-arid areas are underlain by permeable materials, either deposited in intermontane basins and valleys as a result of erosion of the surrounding highlands (as in the case of many of the desert aquifers of the south-western United States), or present as older, more widespread formations, often consolidated or partly so. Given adequate recharge, these formations can be valuable aquifers, vital to the development of these regions. In some areas recharge does take place, either as the result of infrequent but intense rainfall which can locally satisfy soil-moisture deficits or as a result of more regular rainfall on highlands bordering the desert. In some cases, studies of the isotopes (Chapter 11) present in the water indicate that it infiltrated long before the present day. In the Libyan Desert, for example, there are enormous reserves of water in Tertiary sands. It appears that much of this water infiltrated between about 35 000 and 15 000 years ago. It would take thousands of years for groundwater to percolate to the central Libyan Desert from the most likely present-day recharge area in the Tibesti Mountains to the south-west, so this could be one explanation for the age of the groundwater. There is considerable evidence however that infiltration occurred into the desert itself at periods in the past (including the period between 35 000 and 15 000 years ago) when the climate of the area was colder and wetter than at present. These **pluvial** (rainy) periods were related to the glaciations that affected Europe during the Ice Age.

Although limited recharge may be occurring at the present day in several

(a)

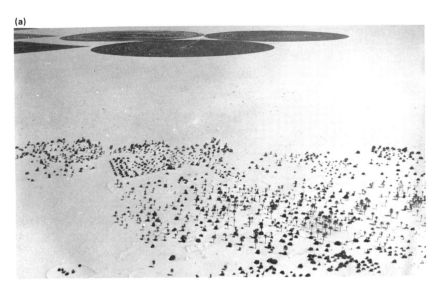

(b)

Figure 8.13 Kufra Oasis, Libya. (a) Aerial view of the desert near Kufra. In the foreground is the vegetation of the Kufra Oasis, which exists as a result of ground-water discharging from the Nubian Sandstone aquifer. The dark circles in the distance are cultivated areas irrigated by sprinklers; each circle is centred on a borehole drawing water from the Nubian Sandstone, and is swept out by a centre pivot irrigator (a sprinkler bar which rotates around the borehole) (b) A centre pivot irrigator and wheat in one of the irrigated areas shown in Figure 8.13a. (Photographs by W.M. Edmunds.)

deserts in the Middle East, most of the fresh water they contain probably infiltrated during pluvial periods.

Because this water entered the aquifers in the past, it is often described as 'fossil water'. Like other groundwater it can be abstracted and exploited, but unlike other groundwater it is not being replenished at the present time. Its exploitation is therefore analogous to that of any other non-renewable mineral resource, such as oil, coal or copper, and for this reason abstraction of 'fossil water' is referred to as 'groundwater mining'.

In Libya, groundwater has been 'mined' for more than 20 years around the Kufra oasis (Figure 8.13). The water is used to irrigate wheat and barley and fodder crops which are fed to sheep that are farmed in the area; the meat is sent to market.

In an even more ambitious project the Libyans are now preparing to pump water through concrete pipelines 4 m in diameter from wells in the Sirte Basin to the coastal area south of Benghazi, a distance of more than 600 km. This is the Great Man-Made River (GMR) Project. In the first phase of the GMR, now virtually complete, the flow will be 2 million m³/day. Later phases will increase this and also bring water from wells in the Fezzan though a separate pipeline to Tripoli (Figure 8.14).

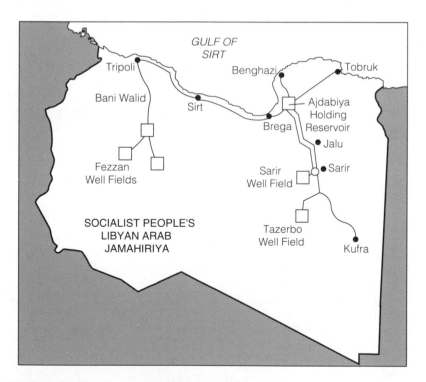

Figure 8.14 Libya's Great Man-Made River. The pipelines that will carry water from well fields in the Libyan Desert to the coastal cities.

There has been some acrimonious debate over the long-term yield of the project and how long the supplies of groundwater will last, but there can be no doubt that the project displays great imagination and engineering skill.

Groundwater quality in desert areas is frequently poorer than in more humid areas. There are a number of reasons for this: predominant among them is the fact that high evaporation rates tend to concentrate soluble salts at the surface, ready to be dissolved by infiltrating water resulting from occasional heavy precipitation. A second major factor is that the limited recharge results in water being 'in residence' in the aquifer for long periods, giving it ample time to dissolve any soluble material present.

However poor its quality may be, however brackish it may taste to those accustomed to the mains water supplies of Europe or North America, this groundwater is vital to desert dwellers. Where groundwater comes to the surface naturally in a desert, the resulting spring or seepage usually leads to growth of vegetation and the formation of an **oasis** (Figure 8.13a). Oases occur for a variety of reasons; sometimes wind has removed the desert sand down to the level of the water table, when the resulting seepage of water and consequent vegetation lead to the stabilizing of the floor of the depression. Other cases arise where confined groundwater is brought to the surface along fault zones or where erosion has removed the overlying confining bed. The quality is not always poor; at the Kufra Oasis in Libya, for example, the groundwater contains less dissolved material than does much of the groundwater supplied for public use in Britain.

'Mining' of groundwater from these arid regions obviously must be carried out with caution. If development of resources is too rapid, quality may be endangered and supplies may be exhausted with little benefit to the inhabitants. Developed carefully, the vast reserves of groundwater in some of these regions can form the basis for agricultural or industrial communities (Figure 8.13) which may be able to continue for many decades – perhaps until the time when water can be brought to them economically from large ocean-desalination plants.

SELECTED REFERENCES

British Geological Survey 1990. *Long-term hydrograph of groundwater levels in the chalk of southern England* (wallchart) Wallingford: British Geological Survey.

British Geological Survey 1992. *Long-term hydrograph of groundwater levels in the Dalton Holme estate well in the Chalk of Yorkshire* (wallchart). Wallingford: British Geological Survey.

Burdon, D.J. 1982. Hydrogeological conditions in the Middle East. *Quarterly Journal of Engineering Geology* **15,** 71–82.

Davis, S.N. and R.J.M. De Wiest 1966. *Hydrogeology.* New York: Wiley. (See especially Ch. 12.)

Doornkamp, J.C. and K.J. Gregory (eds) 1980. *Atlas of drought in Britain 1975–76.* London: Institute of British Geographers.

Hewlett, J.D. and A.R. Hibbert 1967. Factors affecting the response of small watersheds to precipitation in humid areas. In: *Symposium on forest hydrology*, (eds W.E. Sopper and H.W. Lull). Oxford: Pergamon.

Horton, R.E. 1933. The role of infiltration in the hydrologic cycle. *Transactions of the American Geophysical Union* **14**, 446–60.

Houghton, J.T., B.A. Callander and S.K. Varney (eds) 1992. *Climate change 1992: the supplementary report to the IPCC scientific assessment*. Cambridge: Cambridge University Press.

Houghton, J.T., G.J. Jenkins and J.J. Ephraums (eds) 1990. *Climate change: the IPCC scientific assessment*. Cambridge: Cambridge University Press.

Hubbert, M.K. 1940. The theory of ground-water motion. *Journal of Geology* **48**, 785–944.

Ineson, J. and R.A. Downing 1964. The groundwater component of river discharge and its relationship to hydrogeology. *Journal of the Institution of Water Engineers* **18**, 519–41.

Ineson, J. and R.A. Downing 1965. Some hydrogeological factors in permeable catchment studies. *Journal of the Institution of Water Engineers* **19**, 59–80.

Marsh, T.J. and M.L. Lees 1985. *The 1984 drought* (*Hydrological data*: *UK* occasional report). Wallingford: Institute of Hydrology.

Marsh, T.J. and R.A. Monkhouse 1991. A year of hydrological extremes. *Weather* **46**, 365–76.

Marsh, T.J. and R.A. Monkhouse 1993. Drought in the United Kingdom, 1988–92. *Weather* **48**, 15–22.

Marsh, T.J., R.A. Monkhouse, N.W. Arnell, M.L. Lees and N.S. Reynard 1994. *The 1988–92 drought* (*Hydrological data*: *UK* occasional report). Wallingford: Institute of Hydrology.

Parry, M.L. *et al.* 1991. *The potential effects of climate change in the United Kingdom*. London: HMSO.

Rodda, J.C., R.A. Downing and F.M. Law 1976. *Systematic hydrology*. London: Newnes-Butterworth. (See especially Ch. 6.)

Ward, R.C. 1984. On the response to precipitation of headwater streams in humid areas. *Journal of Hydrology* **74**, 171–89.

Ward, R.C. and M. Robinson 1990. *Principles of hydrology*, 3rd edn. Maidenhead: McGraw-Hill. (See especially Ch. 8.)

Wright, E.P., A.C. Benfield, W.M. Edmunds and R. Kitching 1982. Hydrogeology of the Kufra and Sirte basins, eastern Libya. *Quarterly Journal of Engineering Geology* **15**, 83–103.

Additional information is contained in annual publications such as the *Hydrological data: UK* series of yearbooks produced by the Institute of Hydrology and British Geological Survey (Wallingford) and the *National Water Summary* series produced by the US Geological Survey in its Water-Supply Paper series.

9 | Water wells

We do not know where, when or how the first well was sunk. If you had been one of the earliest members of the human race, your water supply would probably have been a river or a spring. If, in some period of dry weather, the spring had gradually ceased to flow or the river had begun to dry up, you would not have been able to explain the facts in terms of a falling water table, but perhaps it would have occurred to you to dig a hollow near the spring; or, noticing that parts of the river bed remained damp, you might have been inspired to dig a hole to reach the water which you would have found just below the surface.

When the flow of the river increased again your hole would have been lost, so that next time the river dried up you would have had to start from the beginning. After a few years of this you may have discovered that it was not necessary to dig the hole actually in the bed of the river; it could be dug a little way from the main channel, where it would still yield water but would not be so readily damaged by the river in flood. In places like the Sudan, the local semi-nomadic people still dig water holes in the beds of rivers that are ephemeral or subject to 'flash' floods; they mark their positions so that they can be found again after the floods have subsided. More permanent water holes are covered with branches to protect them from the floods.

However it began, the digging of wells was an established fact by Old Testament times (Genesis 26). Many Middle Eastern wells were shallow, tapping groundwater a few metres down in wadis or depressions, but some went much deeper. Joseph's Well, near Cairo, dating from the 17th century BC, was sunk to a depth of 90 m in consolidated rock, with a pathway around the sides allowing donkeys to go more than halfway down.

Another ancient Middle Eastern arrangement for collecting groundwater was the 'qanat' or 'kanat'. Kanats appear to have originated in Armenia about 3000 years ago; they were horizontal or gently sloping galleries, up to 30 km long, which intersected groundwater in alluvial-fan material on the sides of mountains and conveyed it, underground, to discharge on the arid plains. Vertical shafts at intervals provided access and ventilation for the workmen digging the tunnel. Kanats were constructed in suitable areas throughout the Middle East, and are still in use today under a variety of names.

The Ancient Chinese developed techniques for drilling, rather than digging, and by sustaining a slow rate of progress for several years could drill boreholes to remarkable depths – some accounts claim as deep as 1500 m. These deep holes were drilled to obtain brine rather than fresh water.

In Europe, early wells were dug shafts. The discovery of overflowing or 'artesian' conditions in Artois and other areas of France and Belgium, and in England, led to the development of borehole-drilling methods to tap these deeper supplies; for a time the term 'artesian well' was used for any well that was bored or drilled, as opposed to those which were excavated.

I am going to follow a convention of referring to wells which are dug or excavated as **shafts** and those which are drilled as **boreholes**. Other distinctions are possible, based on diameter or purpose; all are sensible, but the one I have chosen is convenient for present purposes.

For many people in Britain the word 'well' probably conjures up a picture of a shaft, a metre or so in diameter, lined with stone blocks or bricks, and topped by some form of windlass by means of which a bucket can be lowered on a rope and hauled up again full of water. The well will be surrounded by a wall to prevent animals or children from falling into it, and will probably be of great antiquity and little present-day relevance. The term 'oil well' on the other hand would probably invoke an image of a hole drilled deep into the Earth by modern machinery of great complexity and expense; such is the erroneous relative status of the oil and water industries in the public mind! This arises not from innate ignorance on the part of the public, but because the news media present the oil industry as dynamic and even glamorous; they rarely present the water industry at all unless it is in trouble.

PUMPS

However the well is constructed (and more of this in a while), water must be brought to the surface by artificial means unless it overflows naturally. The bucket and rope may have been adequate for a cottage supply, but most modern wells use pumps. A hundred years ago, most pumps in water wells operated on a reciprocating-piston principle. The upward stroke of a piston in a vertical cylinder drew water into the cylinder, where a simple valve arrangement prevented it from being expelled as the piston moved back down; the next upward stroke forced this water from the cylinder and up into the delivery pipe leading to the surface. Small versions of these pumps could be worked by hand, and they can still be seen in some English villages; such handpumps provide the basic water supply in the rural areas of some developing countries like Bangladesh. Larger piston pumps were powered

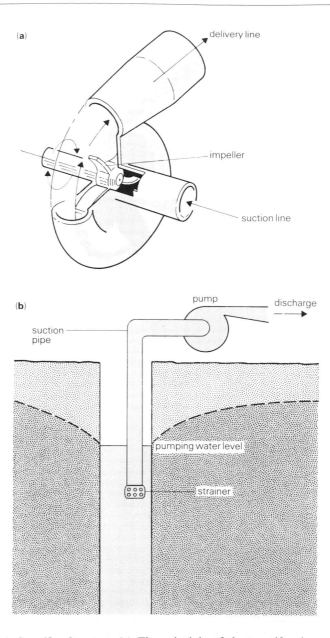

Figure 9.1 Centrifugal pumps. (a) The principle of the centrifugal pump. (b) A centrifugal pump being used to pump water from a shaft. When pumping from a borehole, the suction pipe is often attached direct to the top of the borehole casing. In either case, the vertical distance between the pumping water level and the pump is limited to a maximum of about 7 m.

by steam engines in Victorian English pumping stations, and a few still survive in operating condition.

With the advent of high-speed rotary power units such as electric motors, pumps operating on a rotary principle came into their own. The most important category of these pumps is the **centrifugal** type. In a centrifugal pump, a wheel with vanes – called an **impeller** – rotates causing water inside the pump to move around and outwards (Figure 9.1a). This results in an increase of pressure at the outer wall of the pump and a decrease near the centre of the impeller. Water is thus drawn through the pump from the centre to the edge, where it leaves through a delivery pipe. The pump is in effect transferring energy from the motor to the water.

Centrifugal pumps can be mounted at the surface near a well, with the inlet or suction pipe running down the well to below the water level. The operation of the pump reduces the pressure in the suction pipe (Figure 9.1b) below atmospheric pressure. Because the atmosphere is exerting its full pressure on the surface of the water in the well, water is forced up the suction pipe. If the pump could reduce the pressure at its suction to zero – in other words, create a perfect vacuum – then the atmospheric pressure could raise the water to a height of about 10 m above the water level in the well. In practice, because of mechanical limitations and the possibility of pockets of water vapour (called 'cavities') forming in the water, these pumps cannot operate when the water level is more than about 7 m below the pump.

Where greater lifts are required, the problem is usually solved by putting the centrifugal pump under water with its impeller axis vertical. The pump can then be driven by a vertical shaft running down the well and turned by a motor at the surface; more commonly nowadays a waterproof electric motor is installed beneath the pump (where it is cooled by the water flowing past it) and drives it direct. This type of pump (Figure 9.2) is called an **electric-submersible**. If the water has to be raised more than about 10 m from pumping water level to the surface, several impellers are used, one above the other, each imparting energy to the water. Each impeller is called a **stage**. Five to ten stages are most common, but 20 or more stages can be used for very large lifts. Because the impeller design for these pumps is usually of a type called a turbine, these deep-well pumps are often referred to indiscriminately as **turbine pumps**.

One other form of pumping that deserves mention is **air-lift pumping**. In this method (Figure 9.3) air from a compressor is injected into a delivery main in the well and bubbles up through the water. The resulting frothy mixture of air and water is less dense than the water in the well, so the column of froth in the delivery main must increase in length if the pressure at its base is to be the same as the pressure of the water in the well. If the column of froth could become long enough, the pressures would be equal. If the system is correctly designed, however, the air–water mixture will overflow at the surface before this equilibration can take place; the pressure (and hence the static head) at the bottom of the delivery main will always

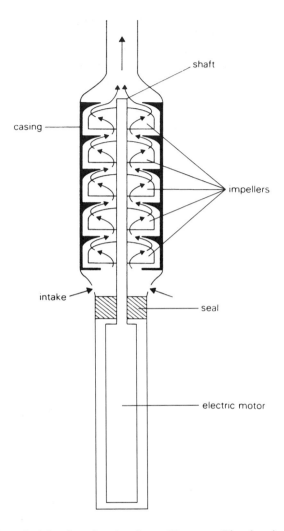

Figure 9.2 The principle of an electric-submersible pump. The electric motor rotates the shaft which carries the impellers. Each impeller in effect operates as a centrifugal pump, drawing water in at its centre and accelerating it outwards and upwards. A seal prevents water from entering the motor.

remain less than that in the well, and water will flow from the well into the delivery main. This water in turn will be converted into froth and the process will continue. This discharge causes a drawdown in the well which in turn causes water to flow into the well from the aquifer.

Air-lift pumping is inefficient and therefore uneconomical for long-term use. It is ideal for pumping a well for a short period, perhaps immediately after its completion, or to save the expense of purchasing a suitably sized pump. It is sometimes used as a long-term method when the water is

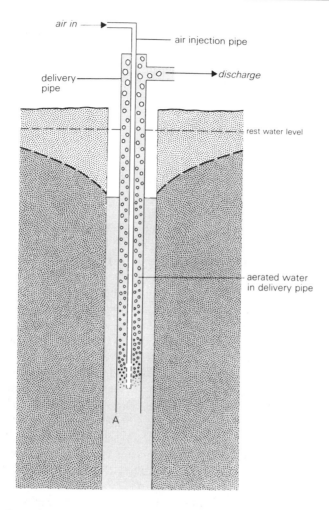

Figure 9.3 The principle of air-lift pumping. The pressure at A is less inside the delivery pipe than outside, because of the low density of the aerated water. Water therefore flows into the delivery pipe from the well.

corrosive, or when it contains sand which would damage the moving parts of a conventional pump.

WELL DESIGN

In consolidated rocks that will stand without support, a well can be very simple, and 'design' hardly enters into the matter. Unconsolidated sediments such as sand or gravel – which are often good aquifers – present a problem, as anyone who has tried to dig a hole on a beach will know; the sand, particularly where it is saturated with water, keeps collapsing into the hole. Above the water table in a well this problem can be overcome. Shafts can be lined with masonry, with pre-cast concrete rings, or with concrete poured in place. Boreholes are usually supported by inserting lengths of pipe, called **casing** or **lining tubes**, of diameter slightly smaller than the drilled diameter of the borehole. The space between the outside of the casing and the borehole wall is usually filled with a thin concrete called **cement grout**. In addition to providing support, this prevents dirty or polluted surface water or soil water from running into the well outside the casing after heavy rain; wells in consolidated rocks are usually given this **sanitary protection** for a few metres below the surface.

Below the water table the insertion of unperforated linings would support the surrounding aquifer but would also prevent the entry of water – not a desirable state of affairs! The well-sinkers who dug some of Britain's older wells got around this problem neatly, if laboriously, by constructing a lining of stones without mortar, like a dry-stone wall. The technique used in modern boreholes is to insert a lining tube that is perforated in such a way as to permit groundwater to flow from the aquifer into the well and at the same time to prevent aquifer particles such as sand grains from entering the well.

The choice of lining will depend on the aquifer. If the aquifer is consolidated but fractured, requiring support to prevent large blocks of rock from falling into the well, then casing with large circular perforations or crude slots will be adequate. If on the other hand the aquifer is loose sand, then a special lining called a **screen** is used. Screens have fine slots, and the appropriate slot size must be specified to suit the particular aquifer. Samples of the aquifer are passed through a series of sieves to determine the sizes of the sand particles, and the screen is chosen so that the slots are as large as possible (to permit the water to flow through the screen with minimum head loss) while being small enough to prevent the aquifer particles from moving into the well. Some of the finest particles will enter the well when it is first pumped – a process called **development** – but thereafter the well should produce sand-free water.

It sometimes happens that the aquifer is composed of particles which are so fine that to exclude them the screen slots would have to be impractically narrow. In such cases coarser slots are used, and an envelope of coarse sand or gravel, called a **filter pack**, is poured into the annular space between the aquifer and the screen. The filter pack retains the aquifer particles outside its outer boundary, and the screen slot size is chosen to prevent the filter pack itself from passing through the slots. A filter pack theoretically needs to be only a few grains thick to function effectively, but it is generally impossible to place sand or gravel in an annulus less than about 60 mm to 80 mm wide. Figure 9.4 shows a theoretical design for a well that draws water from three aquifers – a fine sand that requires a filter pack, a coarse sand, and a sandstone that does not need support. Development of the well

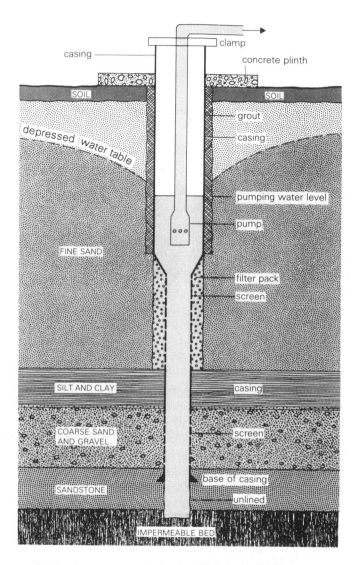

Figure 9.4 Well design. A theoretical well completion in three aquifers: fine sand, requiring a screen and a filter pack; coarse sand and gravel, requiring a screen but no pack; and a consolidated sandstone, requiring no support.

causes removal of the finer particles from the coarse-sand aquifer, leading to the formation of a coarser zone – sometimes called a natural filter pack – immediately around the screen.

THE FILTER PACK

There have been many studies of relationships between sands and filter packs, and numerous published recommendations regarding the ideal size of grains which should be present in a filter pack in order to support a given aquifer. These generally advise that the grains of the filter pack should be x times larger than the grains of the aquifer, where x is a number usually between 5 and 10. Some recommend that the filter pack should contain grains as nearly as possible of one size, while others advise that the pack should contain several grain sizes, i.e. be graded. Most of these recommendations tend to ignore the fact that what really matters is the relationship between the grain sizes of the aquifer and the void or pore sizes in the filter pack; the latter must prevent the aquifer particles from moving through the pack while permitting water to pass through with minimum head loss. So long as this is achieved, it matters very little whether the pack is uniform or graded.

DESIGNING FOR MAXIMUM WELL EFFICIENCY

It sometimes surprises people to learn that wells in a particular area are several tens or even hundreds of metres deep, when the water table or potentiometric surface is only a few metres below ground level. There may be several reasons for this. In the case of a confined aquifer the well must obviously penetrate the full thickness of the confining layer to encounter water, even though the water may then rise to the ground surface. But even in an unconfined aquifer, it is not sufficient for a well just to reach water. If a borehole is drilled just to the water table, then the moment it is pumped – assuming that there is sufficient depth of water for the pump to operate – the water table will be lowered and the well will go dry. So the well must have an allowance for drawdown, which is why in Figure 9.4 the top of the screen is a long way below the water table – there is no point in putting expensive screen and filter pack in the cone of depression. The well need not penetrate the full thickness of the aquifer, but the well designer cannot expect to have the benefit of the total transmissivity of the aquifer if the well penetrates only a fraction of the aquifer thickness.

A related point is that the well must provide an adequate intake area – the area through which water flows into the well. This open area depends on the diameter of the well in addition to the thickness of aquifer it penetrates. A common statement is that doubling the diameter of a well will decrease the drawdown by only about 10% for the same yield – in other words, it is a waste of money to drill a large-diameter well. The statement is correct as far as the aquifer-loss component of the drawdown is concerned, but can be misleading when the well-loss component is also considered. The importance of intake area can be visualized most clearly in the case of a well completed with a screen; if water is flowing into the well at Q m³/s and the total area of the screen slots or perforations is A m², then the entrance speed v of the water through the openings is given by

$$v = Q/A.$$

Experience suggests that v should not exceed 0.03 m/s (although some experiments have suggested that this value is too conservative), or too much energy will be wasted as a result of turbulence as the water enters the well and turns to flow towards the pump. Given this limitation the designer, knowing the discharge Q that is required from the well, can calculate the area of slots needed. This area depends on three factors: the length l and the radius r_w of the screened interval control the total surface area of the cylindrical lining ($2\pi r_w l$); the open area depends on what percentage of the surface area of the screen consists of slots. The most efficient screens have as much as 50% open area; these screens are expensive but may more than justify their extra unit cost by allowing the designer to use a shorter length of screen or one of smaller diameter.

In consolidated rocks, for which linings are not required, the need for adequate intake area to reduce well losses still exists but the problem is usually less than in rocks for which screens *are* required. One aspect in which diameter is important in all aquifers, however, is in accommodating the pump. The upper part of the well, which serves to accommodate the pump (Figure 9.4), is frequently of larger diameter than the lower or intake portion. Because most of the upper section is within the cone of depression, it is lined with plain casing, which is cheaper than perforated casing or screen. This section is kept as shallow as possible, to reduce the expense of drilling and lining this large-diameter hole. As submersible pumps become slimmer, this expense can be reduced further.

DRILLING METHODS

Designing a well is one thing; drilling and completing it to that design is another. Not the least of the problems is the conflict between theory and practice – the theoretical knowledge of the hydrogeologist or engineer who knows what he or she wants, and the practical knowledge of the driller who knows what is possible with the equipment available.

If you want to make a hole in a wall you have a choice of two methods: you can use a rotary bit in an electric drill, or you can use a hammer and chisel. The first is quick, but requires relatively expensive and sophisticated equipment; the second uses equipment that is simple and cheap, but it tends to be time-consuming. The same essential techniques, with the same basic advantages and disadvantages, are available for drilling boreholes in the **rotary** and **percussion** methods.

In the percussion method a heavy drill bit called a **chisel** on the end of a wire rope is alternately raised and dropped to break or disaggregate the rock. The fragments are formed into a slurry either with groundwater or, in unsaturated or impermeable material, by adding a small amount of water to the hole. The slurry is periodically removed using a **bailer**, which is in essence a length of steel pipe with a flap valve at its lower end. As the hole is deepened more cable is fed out from a drum, so that the chisel just strikes the bottom of the hole with each impact.

In its simplest form the technique uses a **drilling string** (the assembly of drilling tools which is lowered into the hole) consisting of little more than the chisel and a rope socket which attaches it to the end of the cable. The percussive or 'spudding' action is provided by the driller alternately engaging and disengaging the clutch on the cable drum, so that the bit is lifted about a half to one metre and then dropped again. This simple technique is widely used for drilling site-investigation boreholes.

The larger percussion rigs used to drill water wells have the spudding action provided by an oscillating beam called the **walking beam**, which is pivoted at one end and which has a sheave at its free end beneath which the cable passes (Figure 9.5). The rocking motion of this beam alternately raises and drops the drilling bit. The bailer is carried on an additional line called the **sand line**. The drilling string comprises the bit or chisel; a sinker bar (in North America this is called a drillstem) to add weight and length and so help to drill a straight and vertical hole; a sliding link called the 'jars' to help in freeing the bit if it gets stuck; and the rope socket. The whole assembly may vary in diameter from 100 mm to a metre or more, and can weigh well over a tonne. Progress is usually good in consolidated rocks but in loose sand the borehole walls must be supported by temporary casing, which is withdrawn once the permanent casing or screen has been installed; these operations can slow down the average rate of progress.

The drilling industry's version of the electric drill is the rotary drilling rig (Figure 9.6). The bit on this machine rotates at the lower end of a hollow steel tube, which is composed of lengths of **drill pipe** screwed together. **Drilling fluid** – originally a mixture of water and clay known as 'drilling mud' but nowadays usually a polymer – is continuously pumped down the drill pipe and through the bit. Its function is to cool and lubricate the bit, and to remove the rock fragments (cuttings) produced as the hole is deepened, by carrying them to the surface up the annular space between the drill pipe and the borehole wall. The pressure of the fluid also supports the borehole wall in unconsolidated material so that temporary casing is only rarely needed. In permeable rocks the liquid part of the drilling mud seeps from the borehole into the formation, leaving a **filter cake** of mud solids on the borehole wall which helps to prevent further fluid loss. At the surface

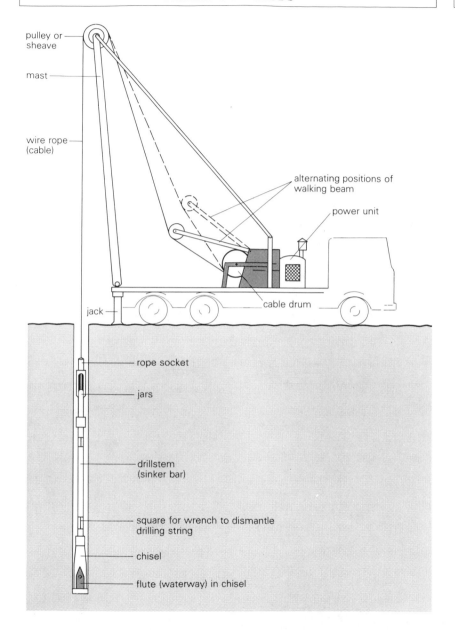

pulley or sheave

mast

wire rope (cable)

alternating positions of walking beam

power unit

jack

cable drum

rope socket

jars

drillstem (sinker bar)

square for wrench to dismantle drilling string

chisel

flute (waterway) in chisel

Figure 9.5 Percussion drilling. The essential parts of a truck-mounted cable-tool (percussion) drilling rig. The bailer and sand line are omitted for clarity.

the mud flows into a settling tank or pit where the cuttings are deposited before the mud is pumped back down the drill pipe.

The rotary motion is applied to the drilling string at the **rotary table**, which is a rotating circular plate with a square hole at its centre, turned

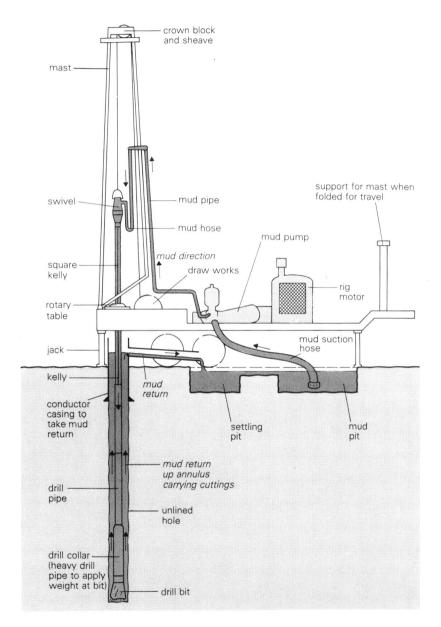

crown block
and sheave

mast

swivel — mud pipe

mud hose

support for mast when
folded for travel

mud direction

mud pump

square
kelly

draw works

rotary
table

rig
motor

jack

mud suction
hose

kelly

*mud
return*

conductor
casing to
take mud
return

settling
pit

mud
pit

drill
pipe

*mud return
up annulus
carrying cuttings*

unlined
hole

drill collar
(heavy drill
pipe to apply
weight at bit)

drill bit

Figure 9.6 Rotary drilling. The main features of a trailer-mounted rotary drilling rig.

by the rig engine. The top length of the drilling string is a special length of drill pipe, of square cross-section, called the **kelly**. This fits in the rotary table and is rotated by it. The weight of the drilling string is partly taken by a wire rope from the rig draw works (in essence, a powerful winch), and the kelly is allowed to slide down through the rotary table as the hole is deepened. When the full length of the kelly has slid through the rotary table in this way, the drilling string is pulled up by an amount equal to one length of drill pipe and the kelly is unscrewed. An extra length of drill pipe is then added before the kelly is reconnected, the drilling string being held meanwhile by steel wedges called 'slips' which fit into the rotary table. The drilling fluid is pumped down the kelly through a hose and a swivel arrangement which permits the mud to enter the kelly while the kelly is rotating.

Rotary rigs range in size and complexity from units capable of drilling holes about 100 m deep, which are usually mounted on tractors or lorry trailers, to the giant machines with masts about 60 m or 70 m high which are used for drilling oil and gas wells. Tall masts permit the use of long individual lengths of drill pipe, thus reducing the time spent in adding or removing lengths. They similarly speed up the process of installing casing. Any rig, whether rotary or percussion, must have the lifting ability to handle the long and heavy total lengths of drill pipe or casing which may have to be lowered into, or pulled from, the well at total depth; this lifting ability is usually what decides the depth of well that can be drilled by a particular rig.

Variations on the basic rotary method include the top-drive rig, in which the rotary power is transmitted directly to the top of the drill pipe by a hydraulic turbine which is lowered down the mast by the draw works as drilling progresses. Another variation uses reverse circulation, in which the drilling fluid (which in this case is usually water) is pumped down the annulus and returns up the drill pipe, which is of large diameter. This method is especially suitable for drilling large-diameter wells in gravels, as quite large cobbles can be brought up the drill pipe without being broken up.

Some rotary rigs use compressed air as the drilling fluid. Another use of compressed air is the method known as **downhole hammer** drilling. For this a small rotary rig is used, but the drilling action is essentially percussive. The drill pipe carries a tool rather like a roadworker's pneumatic drill: compressed air pumped down the pipe operates the cutting tool and carries the cuttings to the surface. The technique is especially suited to hard formations such as basalts; like the air-flush rotary method, its disadvantage is that most compressors are unable to provide air at sufficient pressure to operate the downhole tool and overcome a large head of water. It is therefore not possible to drill to any great depth below the water table with this method, unless the permeability is so low that the air-lift action of the returning air is able to keep the hole effectively pumped dry.

A relatively recent innovation is the downhole motor. In this drilling method the drill pipe remains stationary, serving merely to convey drilling fluid to a turbine which sits at the bottom of the drilling string, coupled directly to the bit. This method was extensively developed in the former Soviet Union. It is gaining popularity in oil-well drilling, largely because it makes possible the drilling of non-vertical holes with precise control over the hole orientation. It has not yet been applied to the drilling of conventional water wells, but it is likely that this application will come.

All of the drilling methods mentioned above have advantages and disadvantages, and no single method is superior to the others for all applications. The percussion method uses simple equipment, requires little water, and makes good progress in most formations except loose sands (which often 'run' (p. 213) into the hole faster than they can be removed by the bailer). A disadvantage is that in unconsolidated formations temporary casing must be installed and subsequently removed; in general, the rate of penetration is slower than that of rotary rigs in similar circumstances.

Conventional rotary drilling is suitable for most formations except those containing large pebbles or nodules in a loose matrix; the pebbles rotate with the bit instead of breaking up, and may deflect the drilling string away from the vertical. Formations containing large open fissures cause problems with **lost circulation**; the drilling fluid flows into the fissures instead of returning to the surface, and fibrous material must be added to the fluid to bridge the fissures. Rotary drilling is fast and the presence of the drilling mud removes the need for temporary casing, but the equipment is more expensive and more complicated, the filter cake may subsequently inhibit the flow of water into the well, and large quantities of water may be needed for drilling.

In some parts of the world shafts are still excavated. Their construction generally requires that people work inside the shaft, which can be dangerous unless adequate – and usually expensive – safety precautions are taken, such as shoring-up the sides to prevent collapse. These precautions and the labour-intensive nature of the work add greatly to the cost, so the method is rarely used except in countries where labour is cheap and safety standards are, unhappily, low. A major disadvantage is that digging usually cannot continue far below the water table, so construction must be undertaken at a time when the water table is at its lowest level – otherwise there is a continual likelihood of the well going dry. Shafts are usually dug in relatively impermeable material, where their large diameters make for useful amounts of storage; unfortunately, the large diameters may also allow their contents to be easily contaminated.

SAMPLING AND CORING

So far we have talked about drilling boreholes in terms of obtaining water, with the emphasis on the hole in the ground as the end product. Sometimes, particularly in the exploration stage of a project, it is the ground that is of interest rather than the hole. These are the times when samples of the aquifer and possibly of any confining beds must be obtained – perhaps for sieve analysis so that correct screens can be ordered or filter packs prepared, or perhaps for laboratory determination of permeability or some other property.

Any drilling method results in fragments of rock being brought to the surface, but the condition and value of the fragments varies with the method. The bailer of a percussion rig brings to the surface a sample whose depth of origin is usually known to within about a metre. The drilling fluid of a rotary rig takes a finite time to bring the cuttings to the surface – time that increases as the hole gets deeper and which may sometimes be over an hour. In that time the hole may have been deepened by 10 m or more, so that the cuttings arriving at the surface are from some way above the current bottom of the hole; allowance has to be made for this, and the cuttings have to be interpreted with caution. Drilling bits are designed to break up and grind away the rock as efficiently and as quickly as possible, so the fragments are often very small; the dust that returns to the surface from a downhole hammer, in particular, is virtually useless to the geologist.

If a relatively undisturbed sample from a precisely known depth is required, the method called rotary **coring** is usually employed. Nowadays this technique makes use of a double-tube **core barrel** (Figure 9.7), which is run into the hole on drill pipe. The outer tube, which rotates, carries a diamond or tungsten-carbide cutting edge, which cuts an annular space around a central core of rock which is typically between 50 mm and 150 mm in diameter. As drilling progresses, the inner tube of the core barrel slides down over the core. The inner tube remains stationary and protects the core from the drilling fluid, which passes between the two tubes and out through special ports in the cutting edge or **crown**. Core barrels on the small rigs used for drilling water wells are usually about 3 m long; when this distance has been drilled and the barrel is filled, the barrel is withdrawn from the hole, the core being retained by a toothed **core catcher** at the lower end of the inner tube. The core is extruded from the barrel at the surface.

The need to withdraw the barrel and all the drill pipe each time the length of the barrel has been drilled means that core drilling takes much longer than conventional rotary drilling. It is therefore more expensive, so cores are usually taken only in exploratory holes, and then only at selected intervals. The method cannot be used in unconsolidated or very broken strata.

Primitive forms of core drilling, in which the core barrel was little more than a tube with a cutting edge, have been used for centuries. In Victorian times many wells were drilled in English Permo-Triassic sandstones using these simple core barrels; steel balls known as 'chilled steel shot' were dropped into the boreholes to aid the cutting action of the barrel. The resulting cores, which were sometimes a metre in diameter, were often dumped outside the pumping stations and can still be seen at some.

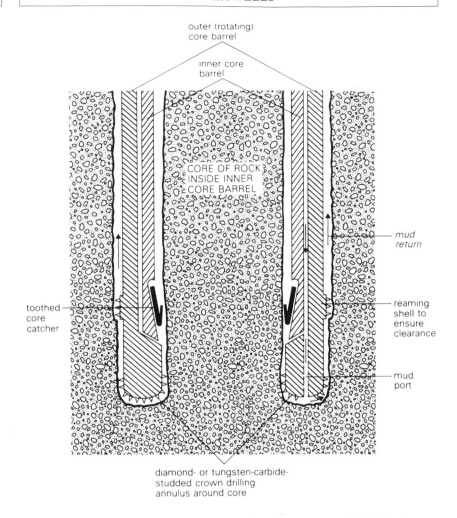

outer (rotating)
core barrel

inner core
barrel

CORE OF ROCK
INSIDE INNER
CORE BARREL

mud
return

toothed
core
catcher

reaming
shell to
ensure
clearance

mud
port

diamond- or tungsten-carbide-
studded crown drilling
annulus around core

Figure 9.7 Rotary core drilling. Schematic section of lower part of a double-tube core barrel. When the barrel is full, the core is broken off by increasing the speed of rotation and pulling up slightly.

Nowadays cores are carefully examined and are usually retained for special tests, including the permeability tests described in the next chapter.

SELECTED REFERENCES

Clayton, C.R.I., N.E. Simons and M.C. Matthews. *Site investigation* London: Granada.

Davis, S.N. and R.J.M. De Wiest 1966. *Hydrogeology*. New York: Wiley. (See especially Ch. 8.)

Driscoll, F.G. 1986. *Groundwater and wells*, 2nd edn. St Paul, Minnesota: Johnson Filtration Systems.

Todd, D.K. 1980. *Groundwater hydrology*, 2nd edn. New York: Wiley. (See especially Ch. 5.)

Vallentine, H.R. 1967. *Water in the service of man*. London: Penguin. (See especially Ch. 10.) Out of print.

10 | Measurements and models

Groundwater is part of the hydrological cycle, but the underground part of that cycle is constrained by geological controls. To study the physical hydrogeology or quantitative groundwater resources of an area three sets of factors have to be considered, requiring three different but interdependent fields of study: **geological**, to investigate the framework in which the groundwater occurs; **hydrological**, to investigate the input and output of water to and from that framework; and **hydraulic**, to investigate the way in which the framework constrains the water movement. None of these three is more important than the others; some hydrogeological investigations have suffered because one aspect was neglected.

HYDROLOGICAL MEASUREMENTS

From the viewpoint of groundwater development, hydrological measurements are needed so that the water balance of the study area can be determined. In its simplest form the water balance can be expressed as:

water entering the area = water leaving the area ± any change in storage.

In more detail this is:

$$\left(\begin{array}{l}\text{precipitation} + \text{streamflow in} + \\ \text{interflow in} + \text{groundwater flow in} + \\ \text{artificial inflow}\end{array}\right) =$$

$$\left(\begin{array}{l}\text{evapotranspiration} + \text{streamflow out} + \\ \text{interflow out} + \text{groundwater flow out} + \\ \text{artificial abstraction of water}\end{array}\right)$$

$$\pm (\text{any change in storage})$$

If a long time period is chosen, beginning and ending at the same time of the year, it is usually safe to assume that storage changes can be disregarded. It is usual to take for the study area a river catchment, so there is no inflow of surface water or interflow from outside the area, although there may be inflow of groundwater if the surface-water and groundwater divides do not coincide (Figure 8.2). Then the water balance becomes:

$$\begin{pmatrix} \text{precipitation} \\ + \text{ groundwater flow in} \\ + \text{ artificial inflow} \end{pmatrix} = \begin{pmatrix} \text{evapotranspiration} \\ + \text{ streamflow in} \\ + \text{ groundwater flow out} \\ + \text{ artificial abstraction} \end{pmatrix}$$

Precipitation can be measured using raingauges, and an average value calculated for the catchment. Evapotranspiration can be calculated from formulae if sufficient climatological information is available, or more direct estimates can be made using lysimeters. A **lysimeter** is an enclosed volume of soil, covered with natural vegetation and maintained under natural conditions, from which the amount of water being evaporated can be measured directly, usually by weighing the isolated soil and vegetation.

Streamflow, including the groundwater and interflow components, can be measured at a gauging station at the point where the river leaves the catchment. Groundwater may leave a catchment as percolation through an aquifer beneath the river; this is sometimes termed **underflow**. If sufficient information is available about the permeability, geometry, and hydraulic gradient, the underflow can be estimated using Darcy's law. Comparison of infiltration estimated from hydrograph analysis (Chapter 8) with estimates based on the difference between evapotranspiration and precipitation will usually indicate whether or not groundwater is entering from another catchment.

Artificial inflows are normally brought in through pipelines, canals or aqueducts, and the amounts involved should usually be known reasonably accurately. The remaining unknown, net artificial abstraction, can be calculated by comparing the withdrawals made at each individual well or river intake with amounts brought into the catchment by pipeline or returned to the ground or drainage system, for example at sewage outfalls.

Sometimes the long-term approach of ignoring changes in storage may be inadequate. If recharge is markedly seasonal and demand for water is greatest in the dry season, then the amount of storage in the catchment becomes of importance. For example, in parts of Asia, where recharge is restricted to the monsoon period and where water is needed for irrigation in the dry season, it may be necessary to compare the precipitation and evapotranspiration, and to calculate soil-moisture deficits and water-table changes, on a monthly or even a daily basis.

Working out the water balance for a catchment is in some ways like checking the financial balance of a household or any other economic unit – it helps us to see where the 'expenditure' is going. If having worked out each item in the equation independently, we put them together and the equation really *does* balance, it gives us some confidence that we know what is happening; if it fails to balance, then we may at least be able to see where more accurate measurements are needed.

Just as checking the household budget enables us to see whether we can afford to spend more or whether we should restrict expenditure, so consideration of the water balance enables us to see whether more use can be made of the catchment's water resources. If there is ample infiltration and ample baseflow to sustain the minimum flow required in the river, for example, then more wells can be drilled to draw on groundwater. However, there are some important points to think about.

Generally speaking, just as we cannot buy something without spending money, so we cannot take water from a catchment without reducing the flow of the stream. Just as an essential decision in monetary management is deciding how much money you can afford to spend without running into problems, so a major decision for the water resources manager is how much water can safely be taken from a catchment, for example for use in public water supply. As always, different people will have their own ideas about this.

It used to be argued that the maximum withdrawal of water from a catchment must not exceed the infiltration into that catchment, so as not to cause a permanent lowering of the potentiometric surface, but arguments of that kind are not readily accepted now. On the one hand, if all the average infiltration is abstracted, then unless water is brought from outside the catchment or sewage effluent is treated and returned to the headwaters of the stream, there will be no water left to form river flow. Environmentalists, anglers and those of us who just enjoy walking along streams would be unhappy about this.

On the other hand, just as it is sometimes necessary to spend more than you earn by taking out a loan to tide you over a difficult period or finance a major purchase, so it may on occasions be prudent to abstract more water from the catchment than is being returned to it. One obvious occasion is during a severe drought (Chapter 8), when nature supplies us with less water than we are used to or need. Then we may be justified in taking out water from storage in the catchment, even though we are lowering the water table to the point where river flow is greatly diminished. Indeed, justified or not, we may have little choice if we want to carry on our way of life.

In drawing on the storage in this way we are effectively taking out a 'loan' from the storage in the catchment. Borrowing is usually acceptable if we are sure we can repay the loan – in other words, if the drought is really a short-term event. If it is not, but really marks a change to a drier climate, then we are not borrowing but 'mining' the resource (Chapter 8, p. 123), something that is usually done only in arid or semi-arid climates.

Deliberate long-term lowering of the water table may be justified in some cases. It can reduce unnecessary baseflow to streams and increase infiltration.

In deciding how much abstraction should be permitted from a catchment, and whether lowering of the potentiometric surface is justified, it obviously helps if water engineers and planners have a thorough understanding of the behaviour of the catchment on which to base their predictions. To achieve this they need to understand not just the hydrology, but the geology and hydraulics of the catchment as well.

GEOLOGICAL MEASUREMENTS

An understanding of the basic geology of an area is vital to a study of its hydrogeology. Several tools are available to help the hydrogeologist achieve that understanding.

The most important of these tools is the geological map, which shows the distribution of various rock formations over an area and permits some understanding of the subsurface structure. Geological maps are available for all but the most remote parts of the world, though the detail and reliability are obviously variable. **Hydrogeological maps**, which combine information on basic geology with data on the hydraulic behaviour of the rocks and their usefulness for water supply, are now available for many countries. In Great Britain, hydrogeological and geological maps are produced by the British Geological Survey. Similar national or regional agencies produce them elsewhere.

Aerial photographs are used in the preparation of maps, and they can be of direct use to the hydrogeologist. For example, where permeable rocks overlie impermeable material there are often springs or seepages (Figure 8.3b); in arid regions in particular, these often show up because of the changes in vegetation that they cause. Seepages along faults and other major fractures or in other discharge areas (Figure 8.3a) can often be detected from the air more easily than from the ground, and the presence of rocks with different permeabilities can influence the surface drainage in a way that is easy to see from the air. Photographs from satellites can now supplement aerial photographs, and films with special emulsions are available to enhance differences in soils and vegetation.

Photography from the air or from space is one of a group of techniques called **remote sensing** which make use of electromagnetic radiation to obtain information about parts of the Earth's surface and its near-surface structure. Usually these techniques produce an image of the ground surface; when the electromagnetic radiation being used is visible light, the image is of course an ordinary photograph. Survey aircraft and satellites now use sensors that produce simultaneously several images of the same area of land, each image being produced by radiation of a different wavelength. These images are processed by computer, and the comparison, combination and enhancement of the images can provide large amounts of information about the vegetation and surface geology of the area. Imagery that uses some wavelengths of infra-red radiation is particularly useful in hydrogeological studies; equipment mounted in an aircraft or satellite builds up an image of the ground that depends on the amount of infra-red energy which the ground surface is reflecting or emitting. By making the survey at a time when solar heating effects are unimportant (usually just before dawn) it is possible to produce an image that depends on the temperature of the ground surface. Because of water's high specific heat (Chapter 4), wet areas tend to stay at a constant temperature while dry rock and soil heat up or cool down more rapidly; this means that seepages can be identified on the infra-red image. Groundwater tends to stay at the same temperature throughout the year, whereas surface water is warmer in summer and cooler in winter; infra-red imagery can use these differences to detect springs discharging into rivers or into the sea.

Mapping and remote sensing provide information about the ground surface. Some idea of subsurface conditions may be inferred from this information, but the only way to find out for certain what is below the ground surface is to drill a borehole. A single borehole provides information at only one place; as a compromise between drilling a network of expensive boreholes and relying on mapping and surface imagery, techniques of **geophysical exploration** are often used. These involve measurement of various physical properties of the ground and are widely used in exploration for metallic ores and for oil. The properties that can be measured include electrical resistivity, local variations in the intensity of the Earth's magnetic field, changes in gravitational acceleration (which is influenced by the thickness and density of the rock formations beneath the point of measurement), and variations in the speed with which sound waves are transmitted through rock layers. Study of the behaviour of sound waves forms the basis of the seismic methods. Some of these techniques can be carried out using sensors mounted in aircraft. This makes possible the surveying of large areas of land in a comparatively short time. Aerial applications of magnetic methods have been especially widespread.

Although many geophysical techniques may be used (cost permitting) to study geological structures during hydrogeological studies, the technique most often used to study the distribution of groundwater is **resistivity surveying**. Most common minerals are poor conductors of electricity, so the ability of a rock layer to conduct electricity depends almost entirely on the amount and conductivity of the water it contains. Measurements made between simple electrodes pushed into the ground can be used to determine the depth to the water table, provided that the geology is straightforward.

For most geophysical surveys, at least one borehole is needed so that the physical measurements can be related to the geology. Geophysical measurements can also be made in boreholes (Figure 10.1); they are then generally known as **well logging** methods. Well logging involves lowering an instrument probe (a **sonde**) into the borehole and making measurements of physical properties of the surrounding rocks or of the borehole itself; a graph (called a **log**) of the property's variation with depth is produced. Properties measured in this way may include electrical resistivity, sonic velocity and various radioactive properties. Measurements of the natural

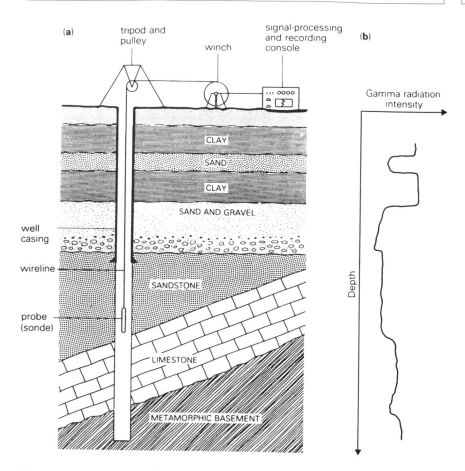

Figure 10.1 Geophysical logging (a) The principles of geophysical logging. (b) A hypothetical gamma log from the strata at (a).

gamma radiation emitted by the rocks around the borehole are useful in correlating strata between boreholes because this radiation depends only on the lithology of the rocks. Most other logging methods measure a response that depends on the porosity and on the properties of the fluid filling the pores. They are therefore useful to the oil industry (which developed most of them) and to hydrogeologists. Figure 10.2 shows a neutron log from a borehole in Permian sandstone and its similarity to a graph of porosity variation with depth; the neutron log responds to the presence of hydrogen nuclei, which are present in the water-filled pores. Porosity measurements can be made only when rock core or cuttings are available, whereas logs can be run in existing boreholes for which samples may no longer be available.

Logs were traditionally recorded directly onto paper or film. Modern equipment converts the signal from the sensor into digital form so that it

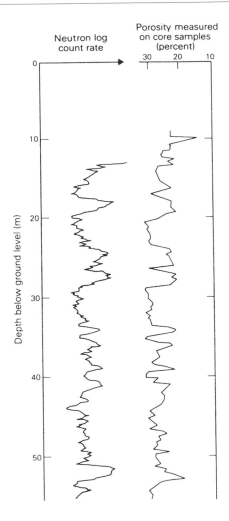

Figure 10.2 Porosity measurement. Comparison of a neutron log (a) with laboratory measurements of porosity (b), made on core samples, from a borehole in Permian sandstone. (Reproduced by permission of the Director, British Geological Survey, neutron log by BPB Industries Ltd.)

can be recorded onto magnetic tape or computer disks. The information can then be processed to remove interference, and logs can be converted to different scales, or combined and compared with other logs to form an even more powerful method of investigation.

Some logging methods cannot be used in boreholes which are lined, and others will work only with certain types of fluid (e.g. drilling mud) present in the borehole. The methods discussed above measure properties of the rock in a zone extending, usually, a few centimetres or tens of centimetres from the borehole, and are referred to as **formation-logging methods**. Other logs can be run to determine the diameter of the borehole, and others are available to measure specific properties of the borehole fluid or its speed of movement. These methods however really belong under the heading of hydraulic measurements, discussed below.

HYDRAULIC MEASUREMENTS

When hydrogeologists are satisfied that they have a reasonable understanding of the geology of an area, they can begin to concentrate on the hydraulic properties of the rocks. They will probably have gained some qualitative knowledge of these properties while studying the geology, but to develop the water resources effectively, quantitative measurements are needed. These measurements often proceed at the same time as water-balance studies, as the two are interdependent.

Aquifer properties

The hydraulic properties of interest (often called the **aquifer properties**) are the porosity of the aquifer or aquifers, which controls how much water is in storage; the storage coefficient, which controls how much of that water can be removed; the transmissivity (or hydraulic conductivity and effective thickness), which governs how readily that water can move through the permeable formations to wells and natural outlets; and the presence and position of **hydraulic boundaries**. Boundaries are in effect limits to the aquifer, because they prevent a cone of depression expanding beyond them. They do this either because they constitute an impermeable barrier (e.g. where an aquifer meets impermeable material at a geological fault, as in Figure 10.3a) or because they provide a source of effectively unlimited recharge. An example of a recharge boundary is a river in hydraulic connection with an aquifer: if a cone of depression expands as far as the river, the river will prevent the cone expanding any further by supplying water to the aquifer (Figure 10.3b). In this way a gaining stream may locally become a losing stream. Both types of boundary cause the cone of depression to become markedly asymmetrical (Figure 10.3).

Other variables that may have to be measured include the hydraulic conductivity of semi-permeable confining beds, through which water may 'leak' to or from the aquifer, and natural hydraulic gradients. These factors are often intimately related to the water-balance studies.

Some of the variables, such as hydraulic gradients and confined storage coefficient, can be measured only 'in the field', i.e. by in-place measurements in the ground (although it is possible to estimate a value of the confined storage coefficient from knowledge of the aquifer thickness, porosity and compressibility). For other variables, such as porosity, permeability and specific yield, a choice of laboratory or field measurements

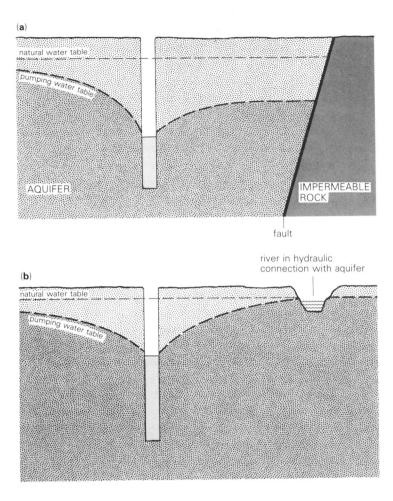

Figure 10.3 Hydraulic boundaries. (a) A fault acting as a barrier boundary. (b) A river acting as a recharge boundary.

may be available. If laboratory methods are to be used, samples of the aquifer must be collected and returned to the laboratory in an unaltered condition. If the aquifer is unconsolidated this is usually impossible. Fissured aquifers also present a problem, since it is rarely possible to collect a representative piece of aquifer containing a fissure with its natural opening preserved. Usually, therefore, it is possible to make laboratory measurements only of the intergranular properties of consolidated materials.

Sampling

The selection of samples requires careful thought. In general, the more samples the better, but a balance must be struck between spending large amounts of time and money studying too many samples, and obtaining unreliable answers by studying too few. Samples must be representative of the bulk of the aquifer, and it is important not to allow bias to creep in when selecting them. If the aquifer is horizontal, then samples representing its full thickness can be collected only from boreholes or cliff sections (Figure 10.4a); inclined formations offer the possibility of sampling the full thickness at the surface (Figure 10.4b) but surface samples are usually affected by weathering which may have altered the properties to be measured. Borehole cores, if carefully sampled and handled, are therefore preferable.

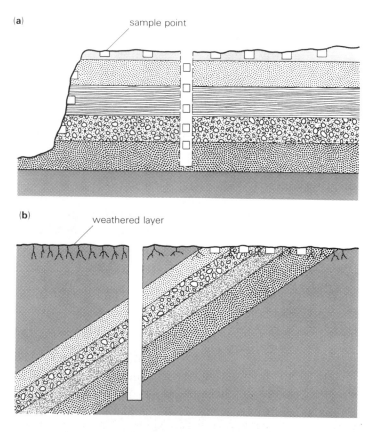

Figure 10.4 Sampling problems. (a) In horizontally bedded strata, samples can be collected from the whole sequence only from boreholes or from cliffs or quarry faces. (b) Where strata are inclined it may be possible to collect samples of the entire aquifer thickness from the outcrop. However, the outcrop material may be weathered and therefore unrepresentative of the aquifer at depth.

Laboratory measurements

Laboratory measurements of porosity can be made in several ways. A simple way which is accurate enough for most purposes is to dry the sample, weigh it, and then evacuate it in a sealed container. A liquid of known density is then introduced into the container so that it covers and saturates the sample. The sample is weighed submerged in the liquid and, still saturated, in air; application of Archimedes' principle allows the bulk volume to be determined. The difference between the dry and saturated weights in air, divided by the specific weight (ρg) of the liquid, gives the pore volume; the porosity can therefore be calculated. In practice, many samples can be evacuated and saturated simultaneously. By achieving controlled artificial drainage of the saturated sample, in a centrifuge for example, an approximate specific yield can also be determined.

Laboratory instruments for measuring permeability are called **permeameters**. In essence a permeameter is simply a device for holding a sample so that a fluid can be passed through it and Darcy's law applied to determine the hydraulic conductivity. Holding a sample of rock so that the fluid passes only through the rock and not through crevices between the rock and sample holder is difficult, and much research has gone into ways of achieving it. When dealing with consolidated aquifers most laboratories cut small cores, typically 25 mm to 50 mm in diameter and length, or cubes of similar size, from the borehole core or outcrop sample. At least two cores are usually taken from each sample, one with its axis parallel to the bedding and one at right angles to the bedding, so that permeabilities can be determined in these two directions. Cubes have the advantage that permeability can be measured in three directions on the same piece of rock by mounting the cube in the permeameter in different orientations.

Ideally, the fluid passed through the sample should be groundwater from that formation, and no other fluid should come into contact with the samples from the time they are part of the aquifer until the time the tests are finished. This is because groundwater is not pure water: it contains dissolved materials in various proportions and concentrations. The groundwater and the aquifer will be in approximate chemical equilibrium; if water with different dissolved material is introduced into the rock, the minerals in the rock may change. This is most likely to happen when clay minerals (present in many sedimentary rocks) are involved; the clay particles may swell or shrink as the water chemistry is changed, thereby altering the dimensions of the pores and hence the rock properties. (Even the action of drying samples for a porosity test can cause large changes, and sophisticated drying techniques may have to be used.) Values of permeability differing by a factor of a hundred or more can be measured on the same rock sample using different test fluids; if natural formation water (or a solution with the same dissolved constituents) is used for all measurements and preparation, the possibility of error is greatly reduced.

When dealing with 'clean' rocks (those containing negligible amounts of clay) the choice of test fluid is less critical, and distilled water or even a gas can be used for permeability measurements. Gas measurements are quick, as small gas flows can be measured accurately, whereas small liquid flows may have to be collected for a long time to obtain a measurable volume; however, allowances must be made for changes in gas volume caused by expansion and for the fact that gas molecules 'slip' through small pore channels more easily than do liquid molecules. Gas techniques are also available for the measurement of porosity.

Laboratory tests offer the possibility of making accurate measurements under carefully controlled conditions on samples of precisely known geometry, but the samples are inevitably small. To test more representative volumes we must leave the laboratory and move 'into the field'.

Field measurements

One of the most effective and frequently used methods of measuring aquifer properties is the **field pumping test**. We saw in Chapter 6 that when water is pumped from a well, a cone of depression is formed in the potentiometric surface. The steepness of the cone depends on the hydraulic gradient, which in turn depends on the pumping rate and on the transmissivity and storage coefficient of the aquifer. The storage coefficient relates the volume of the cone to the total quantity of water pumped out; the smaller the storage coefficient, the larger the cone must be for any given quantity of water abstracted. Knowing these relationships it follows, in the absence of complicating factors, that if we pump water from a well and observe the way the cone of depression expands, then we should be able to deduce the transmissivity and storage coefficient of the aquifer. This is the principle of the pumping test, but the phrase 'in the absence of complicating factors' must be kept in mind.

The usual procedure for a test is that water is pumped from one well – called the **production well** or **pumped well** – at a constant rate, which is carefully controlled and measured. The resulting change in the potentiometric surface is monitored by measuring the change in head in one or more **observation wells** near the pumped well. (Provided that the density of the groundwater is constant, the change in water level in a well is an accurate representation of the change in head.) A possible arrangement of wells for a test is shown in Figure 10.5. The longer pumping continues, the further the cone of depression will expand. At greater pumping rates the hydraulic gradient must be steeper and the cone must therefore be deeper, causing larger drawdowns; however, simply increasing the rate of pumping will not cause the cone to extend further outwards.

In a perfectly confined aquifer (one that has totally impermeable confining beds), the cone of depression will continue expanding until it meets some form of boundary. The growth of the cone will be rapid at first, but will

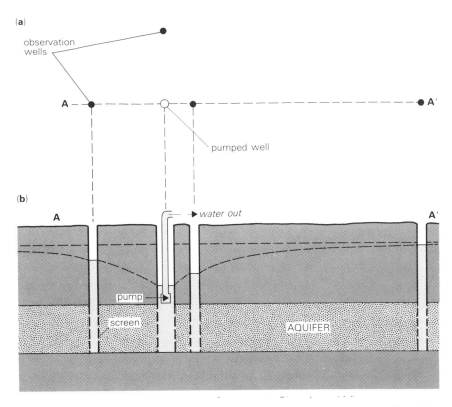

Figure 10.5 Idealized layout for a pumping test. (a) Plan view. (b) Section along line A–A'.

become slower, because each time the radius of the cone is doubled the volume is quadrupled; as the pumping rate is constant, the *volume* of the cone must increase uniformly and its *radius* must therefore grow ever more slowly. The cone must grow at a finite (albeit slow) rate until it intersects some form of recharge.

If the potentiometric surface was initially horizontal, and if the aquifer is perfectly homogeneous and isotropic, the cone of depression will be symmetrical about the pumped well. For these conditions Charles V. Theis, of the United States Geological Survey, produced in 1935 an equation relating the drawdown of the potentiometric surface at any distance from the production well to the transmissivity and storage coefficient of the aquifer and the rate and duration of pumping. Because the cone of depression is assumed to be symmetrical, only one observation well is needed, in theory; however, if only one observation well is used, it is impossible to tell whether or not the required symmetry is present. In practice, it is therefore advisable to use at least two observation wells in addition to the production well; measurements of drawdown are made at frequent intervals in all wells.

The nature of the Theis equation makes it easy to calculate the drawdown when the properties of the aquifer and other factors are known, but difficult to calculate the aquifer properties when the drawdown is known. The usual way of using the equation is therefore to plot a graph of drawdown against time for each well on double logarithmic (log–log) graph paper, and to match these data plots to a 'type-curve' drawn on similar paper. From the relative positions of the graph scales, the transmissivity and the storage coefficient can be determined. If several observation wells are used, data from them all should yield roughly the same values of transmissivity and storage coefficient; if they do not, the cone of depression is not symmetrical and the Theis equation is not strictly valid. If no observation well is available (this sometimes happens in the early stages of a groundwater investigation, or when testing very deep aquifers) then measurements in the production well make it possible to determine the transmissivity but not the storage coefficient.

Theis derived his equation by using an existing solution for an analogous problem in heat flow. It was a tremendous step forward, because until its derivation the commonly used pumping-test formula was one developed in 1906 by G. Thiem in Germany from earlier work by the Frenchman, J. Dupuit; their equation describes the flow of groundwater to a well under equilibrium conditions. These conditions exist only if the aquifer is receiving uniform recharge around the cone of depression, although in practice they are approximately satisfied after long periods of pumping, when the cone of depression is expanding very slowly. Because it assumes that the cone of depression has stopped expanding and water is no longer being taken from storage, the Dupuit–Thiem equation cannot provide a value for the storage coefficient.

The Theis equation therefore ranks as of fundamental importance in hydrogeology. Various approximations allow it to be used graphically without the need for type-curves, and it can also be used with data from the recovery of the potentiometric surface at the end of pumping. Other workers have produced techniques or modifications for use when the aquifer does not conform to the strict requirements of the Theis equation, namely that the aquifer is perfectly confined, has an infinite horizontal extent, is homogeneous with constant transmissivity in space and time, has an initially horizontal potentiometric surface, and releases water from storage instantaneously as the head is reduced; and that the pumped well is of negligible diameter, completely penetrates the aquifer and is pumped at a constant rate.

Theis himself suggested that some of the assumptions were more important than others. For example, if the diameter of the well is large, it takes a finite time for the water in the well to be removed; the effect is seldom important in practice except in dug shafts or in very deep wells, or where the aquifer is of low permeability; in all these cases, the water in storage in the well may be equal to many minutes' worth of pumping. 'Infinite

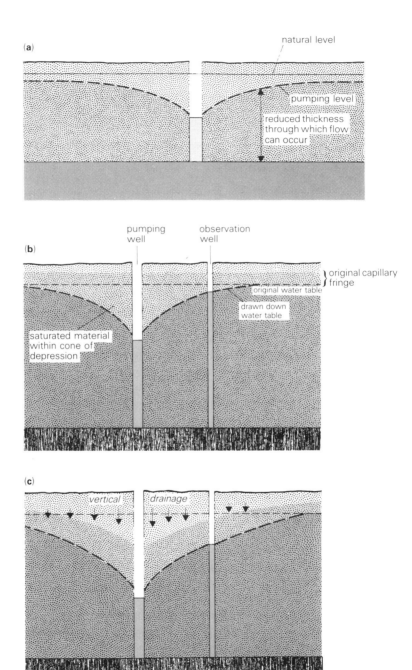

Figure 10.6 The cone of depression in an unconfined aquifer. (See text.)

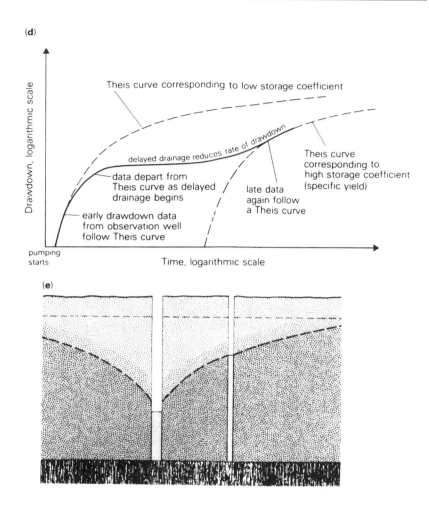

(d)

Drawdown, logarithmic scale

Theis curve corresponding to low storage coefficient

delayed drainage reduces rate of drawdown

data depart from Theis curve as delayed drainage begins

late data again follow a Theis curve

Theis curve corresponding to high storage coefficient (specific yield)

early drawdown data from observation well follow Theis curve

pumping starts

Time, logarithmic scale

(e)

horizontal extent' means, in practice, that the cone of depression never reaches a boundary; if it does, a recharge boundary will reduce the rate of drawdown and a barrier boundary will increase the rate of drawdown.

If the aquifer is not perfectly confined, water may leak through the confining beds into the aquifer in response to the reduction in head caused by pumping. This leakage, like a recharge boundary, will reduce the rate of drawdown by supplying additional water; Mahdi Hantush produced the first equations to deal with this problem.

When water is pumped from a confined aquifer, the aquifer usually remains fully saturated; only the water pressure is reduced. In an unconfined aquifer (Figure 10.6a) withdrawal of water means that the water table is lowered, and the saturated thickness of the aquifer is reduced. This alone can cause problems in analysis; however, as Theis pointed out, if the drawdown is small in relation to the saturated thickness of the aquifer, the resulting change in transmissivity may be small. A more important problem is the way in which the water is released from storage. When pumping starts, the head around the pumping well declines rapidly, as it does in a confined aquifer; the water table, which is the level at which the water is at atmospheric pressure, is therefore lowered. As in a confined aquifer, the first water released comes from elastic storage as the aquifer compacts and the water expands slightly in response to this reduction in pore-water pressure. Initially, however, little or no water drains from the pore space. We therefore have the condition shown in Figure 10.6b; the water table has been lowered to form the cone of depression, but within this cone most of the material is still fully saturated – in effect, we have a thick conical capillary fringe. For a short time after pumping starts, the drawdown curve therefore follows that predicted by the Theis equation (Figure 10.6d).

The thickness of saturated material in the cone of depression is generally too great to remain saturated, however (the capillary columns are too long in relation to their diameters), and so this material begins to drain vertically towards the water table (Figure 10.6c). This drainage has the effect of recharge entering the cone of depression: it causes the rate of drawdown to be less than that predicted by the Theis equation (Figure 10.6d).

As pumping continues the cone of depression expands more slowly, and the drainage of the pores within the cone is able to catch up and keep pace with the growth of the cone (Figure 10.6e). The time–drawdown behaviour of the aquifer (Figure 10.6d) returns to the curve predicted by the Theis equation, but the curve is displaced from the starting point. The complete response of an unconfined aquifer to pumping can thus be thought of as comprising three stages: initially, the response is like that of a confined aquifer, with water being released from compressible storage; in the second stage, the rate of drawdown decreases as vertical drainage takes place from saturated material within the cone of depression; then, as this drainage catches up with the growth of the cone, the drawdown again follows a Theis curve, but one corresponding to a higher storage coefficient (the specific yield). The time for the second stage (which is known as the **delayed yield** stage) to be completed can range from hours to weeks or more; it will be prolonged if drainage is delayed by material of lower vertical permeability or by capillary effects.

The reduction in transmissivity caused by reducing the saturated thickness of the aquifer can cause further complications, as can the vertical flow components within the saturated part of the cone; the response of unconfined aquifers is therefore complicated. The first person to produce equations describing the three-part nature of the response was N.S. Boulton in England; the understanding of unconfined behaviour was greatly advanced by S.P. Neuman in the 1970s. These advances, like many more in pumping-test analysis, relied on the availability of computers to handle the numerical calculations involved for the many different cases. It is worth noting, however, that an essentially correct qualitative description of a cone of depression in a water-table aquifer was given in 1850 by Robert Stephenson, the famous civil engineer (and son of the even more famous 'father of the railway', George Stephenson). In a report on the water supply of Liverpool, Stephenson not only described the shape of a cone of depression, but correctly surmised the way in which a geological fault would act as a barrier to groundwater movement – all this, six years before Darcy published his work.

Most hydrogeologists would agree that the pumping test is generally the most satisfactory way of measuring aquifer properties. Unlike laboratory techniques it tests a significant volume of the aquifer, in place, whether the flow through it is predominantly through pores, through fissures, or through a combination of the two; it does this, in addition, using formation water. When something seems too good to be true, we are right to be suspicious!

To begin with, compared with laboratory measurements, what the field test gains in volume tested it loses in precision and control. No longer is the tested sample of carefully measured size, held in a permeameter; it is an uncertain volume of aquifer, whose thickness and lithology can be precisely ascertained only where expensive boreholes have been drilled. And an aquifer that looks uniform on the basis of information from two or three small-diameter boreholes can contain some large surprises. One of the beauties of the pumping test is that if it continues for long enough the aquifer properties deduced from it will be useful averages of the minor variations that are present; the problem comes in deciding when it has continued for long enough. A test may, by chance, be ended after a 'minor variation' alone has been studied, and this fact might not be discovered until expensive development boreholes have been drilled in the wrong places or to the wrong depths.

If one knew all the variations in the shape of the cone of depression in both space and time, it might be possible to produce a unique and accurate description of the hydraulic variations that the cone has encountered. Unfortunately, we cannot achieve this complete knowledge: the only points at which we know the position of the potentiometric surface are the observation wells. Pumping tests do not therefore avoid the problem of sampling: we are effectively 'sampling' the cone of depression when we make measurements in the observation wells. In many tests, several sets of conditions could produce the same 'sampled' results.

Finally, we have to ask whether the aquifer is undisturbed by the testing. Probably it is not. As we have discussed earlier (Figures 6.10c and 8.1, for example), there are vertical components of flow – and therefore vertical hydraulic gradients – in most aquifers. These gradients can cause the heads at different depths in the aquifer to differ significantly; the presence of layers with low permeability will tend to cause these differences to be increased. Boreholes can provide highly permeable pathways between these points with different heads, creating what in electrical terms would be a 'short circuit' and disturbing the very parameter they are intended to measure.

The combined approach to hydraulic measurements

A cynic might summarize the position by saying that laboratory tests permit accurate measurements to be made on disturbed, unrepresentative material; whereas field methods provide rather inaccurate measurements on what was representative material only before the test boreholes were drilled through it. Even to the least cynical, it may appear by now that aquifers are so variable, and the methods for studying them so imprecise and beset with problems, that the prediction of groundwater behaviour requires a crystal ball rather than a computer. Most hydrogeologists would probably agree, at least in part, with such a conclusion; nevertheless the predictions must be made and reliable crystal balls are hard to find. What should the hydrogeologist do? Field studies are expensive and unlikely to provide a unique interpretation; the apparent precision of results from laboratory methods depends on which samples were chosen, and cannot take account of fissures or aquifer boundaries. If time and money are available, the investigator will probably choose somewhere between these two extremes. Just as an understanding of the hydrogeology of an area involves studying the geology, hydrology and groundwater hydraulics, so the study of the groundwater hydraulics can itself be properly achieved only by employing a variety of techniques.

How far it is necessary or possible to pursue this type of combined investigation will depend on circumstances such as how much is already known, and whether the investigation is for a major scheme involving many wells over a large area or for a single production well. It will depend too on how quickly the results are needed, and on how much money and how many people are available for the work; these last circumstances are usually beyond the control of the hydrogeologist. Scientists can make recommendations, but the decisions are usually made by politicians, and, unfortunately, are sometimes made for political rather than technical reasons. Thus it may happen that large schemes, on which votes or prestige may depend, are rushed through with less time for investigation than smaller ones; those in authority often feel that a job which should take four years should be possible in one year if four times as much money is made available. This sort of attitude overlooks the fact that trained and experienced workers cannot be obtained at the drop of a hat, that exploration programmes must progress in a phased way if money is not to be wasted, and that several years of data collection may be necessary for proper plans to be made.

Leaving aside political and economic constraints, an investigation of the groundwater hydraulics of an area of hundreds of square kilometres might commence with a series of exploratory boreholes, some of which would be core drilled if possible and all of which would be logged. The holes could then be used for measurement of water-level fluctuations and natural hydraulic gradients. Preliminary pumping tests would be carried out to obtain some idea of the transmissivity of the aquifer or aquifers. Also at this stage samples of water would be taken for chemical and biological analysis (Chapter 11). If all the results were encouraging, a further pumping test or tests would be undertaken with observation wells.

At least some of the cores would be subjected to laboratory measurement of permeability, and the results compared with those obtained from the pumping tests. Frequently there will be discrepancies between the two, the presence of fissures commonly resulting in field values higher than the corresponding laboratory values.

Geophysical logging of the borehole fluid column is an excellent way of detecting or verifying the presence of fissures. The temperature and chemical quality of groundwater usually vary slightly from place to place and with depth, even in a single aquifer. The column of water in a borehole usually represents an average of the groundwater in the surrounding aquifer, and at any particular level in the aquifer the properties will usually differ slightly from this average. If water enters the borehole in relatively large quantities from a particular level – because of the presence of a transmissive fissure or of a layer of highly permeable rock – there will thus be a slight change in the temperature and chemical quality of the borehole water at that level. These changes can be detected using sensitive temperature and electrical-conductivity logging equipment, even though the temperature change, for example, may be only 0.01 °C. Figure 10.7a shows an actual example of how an electrical conductivity log indicates a fissure.

Other fluid logging instruments include a variety of **flowmeters**, which measure the speed of flow of water up or down a borehole. The movement may result from natural head differences or be the result of pumping; major changes in speed are associated with levels where water enters or leaves the borehole – again these are fissures or highly permeable layers (Figure 10.7b).

Finally, what is in some ways the simplest technique (though it requires sophisticated equipment) is to look at the borehole wall using an underwater closed-circuit television camera, with a television monitor on the surface. Fissures can be observed directly and their openings and orientation estimated. The results can be recorded on videotape or the screen can be photographed (Figure 7.7).

Combining all these techniques, the hydrogeologist can either obtain a reasonably consistent picture of the aquifer or, at worst, see where further study is needed.

To obtain a quantitative estimate of the contribution which particular layers or fissures make to transmissivity at a site, sections of boreholes can be tested individually by isolating them between inflatable packers (rubber sleeves which are expanded to seal against the borehole wall) and pumping water into or out of the isolated section (Figure 10.8). Injecting water into a test interval can change the aquifer chemistry around the borehole; testing

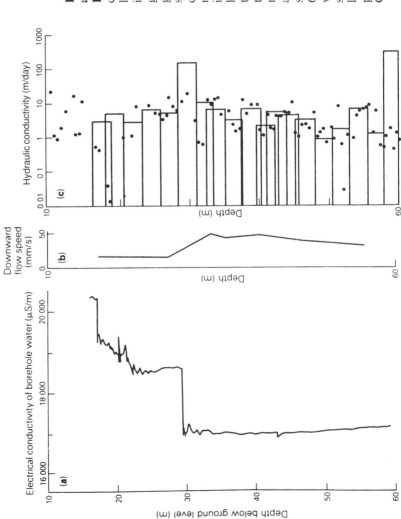

Figure 10.7 Geophysical and permeability measurements in a borehole in Permian sandstone. (a) An electrical conductivity log. (b) A flow velocity log. Both (a) and (b) indicate a major inflow of water at about 29.5 m below ground level. (c) A comparison of intergranular hydraulic conductivity measured in the laboratory, indicated by dots, and total hydraulic conductivity measured using packer tests (see text), indicated by rectangles. (The right-hand end of each rectangle indicates the average hydraulic conductivity for the depth interval represented by the rectangle.) There is generally good agreement between the two sets of measurements except at about 30 m and 60 m where television inspection revealed major fissures – the higher fissure caused the inflow detected by the logs in (a) and (b). (Reproduced by permission of the Director, British Geological Survey.)

by pumping out avoids this, and other, complications, but is more difficult in practice (Figure 10.8b). An example of the results obtained using packer tests, and how they compare with laboratory measurements and flow logs, is shown in Figure 10.7. This shows that consistent results can be obtained if sufficient trouble is taken to collect the data.

MODELS

Having gone to some trouble to collect geological, hydrological and hydraulic information about an area, what do we do with it? Usually the information is needed so that the water resources of the area can be evaluated, calculations made of the rate at which water can be abstracted, the effects of abstraction predicted, and so on. Before these calculations can be made, a **model** of the aquifer is needed.

At its most abstract, a 'model' is a unification of the concepts about the aquifer which the hydrogeologist uses in making his or her* predictions. For example, he may have concluded from his investigation that the aquifer is confined, has high intergranular permeability, and is traversed by numerous fissures which cause local variations in aquifer properties but which, on a regional scale, increase the transmissivity in a uniform way. Also, the aquifer extends well beyond the study area. As a first approximation, he may therefore be justified in treating the formation as an infinite, homogeneous, isotropic confined aquifer to which he can apply the Theis equation to predict drawdown around pumping wells; this would be his conceptual model of the aquifer. If he found that the fissuring was restricted mainly to the upper part of the formation, causing a markedly higher permeability there, he would probably change his idea to a two-layer model. He can use these models, or ideas, as the basis for manual or graphical predictions of, say, drawdown resulting from the simultaneous pumping of several wells.

As the number of wells increases or the aquifer conditions or geometry become more complex, it becomes more difficult and time-consuming to do the calculations, even if the equations can be solved. A model is then needed that is more than just an idea.

One possibility is a *physical* model of the aquifer, which usually takes the form of a **sand-tank model**. In this model the aquifer is represented by a layer or layers of sand in a watertight tank. Impermeable beds or boundaries can be sheets of plastic, and the potentiometric surface is observed in thin tubes which act as observation wells or piezometers. Other tubes serve as production wells. A major problem is that capillary effects tend to be large in relation to the size of the tank, making it difficult to model regional unconfined aquifers; however, sand models have been widely used to study flow near wells.

*Although 'he' is used for brevity here, many hydrogeologists are, of course, women.

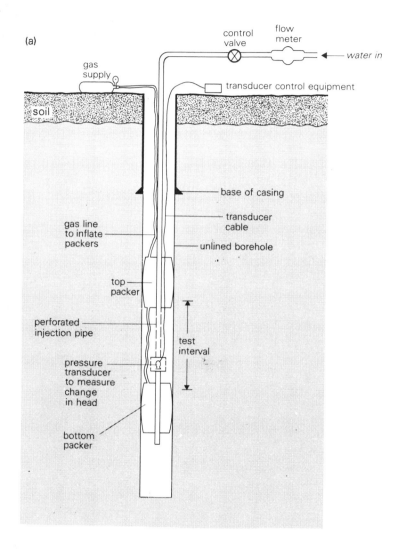

Figure 10.8 Arrangement for testing a borehole interval isolated between packers. (a) Injection test. Water is pumped into the interval at a measured rate, and the resulting head change is measured with the transducer. (The water is omitted for clarity.)

Another type of model, offering more scope for regional simulation, is the **analogue**. This uses the analogy between the laws of groundwater movement and the laws governing the movement of some other physical quantity. One of the most versatile analogues, the **resistance network**, uses the similarity between the flow of groundwater and the flow of electric current. (See the box on p. 50.)

In resistance analogues a model of the aquifer is constructed using

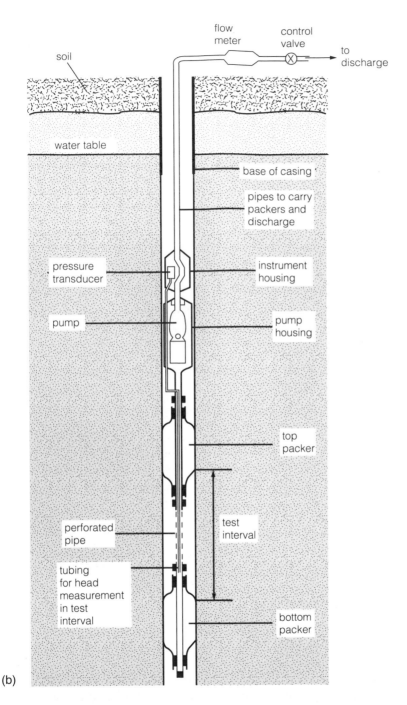

soil

flow meter

control valve

to discharge

water table

base of casing

pipes to carry packers and discharge

pressure transducer

instrument housing

pump

pump housing

top packer

perforated pipe

test interval

tubing for head measurement in test interval

bottom packer

(b)

Figure 10.8 continued. (b) Pumping-out test. In this arrangement the pump is built into a housing connected to the test interval and drawing water from it. A pressure transducer is connected to the test interval to allow measurement of drawdown. If the packers jam in the hole, the pump and transducer can be retrieved. (Gas line, cables, vent tubes and the water in the borehole are omitted for clarity.) (After Price and Williams, 1993.)

electrical circuit components. Usually the models are built on a grid basis with electrical components meeting at junctions called **nodes** which correspond to the intersection of grid lines on a map of the aquifer. The resistor connecting two nodes is chosen to have a resistance that is inversely proportional to the transmissivity of the aquifer between the corresponding nodes on the map. The electrical potential (voltage) at any node represents the head in the aquifer at the corresponding point, and electric current represents flow of water. The model conforms to the shape of the portion of aquifer being considered, and appropriate conditions of current and potential must exist at the boundaries. We saw on page 144 that the water entering a study area must equal the water leaving the study area unless there is a change in the water stored within the area. In summarized form: inflow – outflow = change in storage. If there is no change in storage, inflow = outflow and the system is said to be in **steady state**. When modelling flow through an aquifer under steady-state conditions it is not necessary to know the storage coefficient. If flow is not steady, storage has to be considered; in the model this is achieved by connecting capacitors, which store electricity, to each node. The capacitance of the capacitor is proportional to the storage coefficient of the aquifer at the corresponding point.

The grid is drawn in such a way that the size of the square on the ground is small in relation to the area of aquifer being studied. It is not practical to measure the aquifer properties in every grid square; usually they are measured in a few places and all other values are estimated or interpolated. The model is **calibrated** using historical data; observed values of head are fed into the model as starting conditions, and the model's prediction of the head changes after a suitable interval is compared with what really happened. The analogous 'head' (the potential) is measured with some form of voltmeter. The components used in the model are chosen in such a way that processes which take several years in the aquifer are simulated in the model in seconds.

If the predictions from the model do not agree with the historical records, the resistance or capacitance values probably do not accurately represent the aquifer properties. They must be altered, by changing the components, until agreement is achieved.

Once the behaviour of the model agrees with the behaviour of the aquifer, it can be used to predict the aquifer's response to changing conditions. Pumping can be simulated, for example, by using an electronic signal generator to produce the appropriate voltage fluctuations at the node corresponding to the production well.

Electrical analogues were popular in the 1960s. Their main disadvantage is that they are laborious and time-consuming to construct and modify during calibration, and that they can simulate the behaviour of only one aquifer. They are usually constructed as two-dimensional models, with no simulation of layering or vertical flow. Since the 1970s their place has increasingly been taken by **digital models**, which make use of the increasing power and availability of digital computers; they are in essence computer programs. However, it is worth noting that there are still a few problems for which analogue models are superior, and they can also be valuable teaching tools.

From the mathematical point of view, the simplest type of digital model is the **finite difference** model. This also involves dividing up the aquifer into smaller units; squares are often used for simplicity. As with analogue models, steady-state conditions are generally easier to model than those where the water in storage is changing. Formulae are fed into the computer to calculate the head conditions within each square, or at the intersection of squares, from the heads in adjacent squares and making use of knowledge of the variations in hydraulic conductivity or transmissivity. (Where K or T is low, hydraulic gradients will be steeper and heads will change more rapidly.) At the edges or **boundaries** of the model, the modeller must specify the heads or specify where no flow can take place. Once these boundary values have been set, the model will calculate the heads in all the other squares using the formulae that have been programmed into it. Simple problems can be solved with a spreadsheet package on a personal computer.

Where flow is not steady, the aquifer properties for each square and the initial heads in the squares are fed into the computer, which then calculates the flow that will occur over a short time interval and what the heads will be at the end of that time. It repeats the process as many times as required, taking into account information on recharge and abstraction.

As with analogues, a digital model must be calibrated using historical data before it can be used to make predictions. The advantage of the digital model is that changes found to be necessary during calibration require only a new set of values to be fed into the computer, not physical alterations to the analogue. Similarly, the same model can be used to study different aquifers by changing the aquifer properties and boundary conditions. Standard modelling 'packages' are available which can be used to simulate many problems by the relatively simple expedient of the user entering the boundary conditions, assumed aquifer properties and other data. These packages include models to simulate the behaviour of aquifers of complicated shape and variable lithology, to predict changes in water chemistry, and to calculate the velocity of movement or spread of pollutants.

In the manner of Parkinson's law, models expand to fill the computer storage available. Desk-top – and even portable – computers available today will handle models which, years ago, would have needed a mainframe computer, but modellers keep developing models of greater sophistication which use to the full the storage available on the newest computers. The use of these models can often reveal flaws in existing knowledge of the aquifers. However, discretion has to be exercised in their use; the phrase 'rubbish in, rubbish out' indicates the predictions that can be expected from a model, however sophisticated, which is fed with unreliable data or which is based on incorrect suppositions.

There are perhaps two main dangers. The first is that the writing (as opposed to the everyday use) of models is becoming such a specialized activity that modellers may become divorced from the real-life situations and data that they are attempting to understand and use. The second is that by trying to counter this problem, and put 'user-friendly' models into the hands of working hydrogeologists who are in touch with day-to-day problems, modellers may be giving those hydrogeologists tools that they do not fully understand and may misuse. The tendency for 'field' workers and data interpreters to become separate and specialized is one that has happened in geophysics. It has benefits of efficiency and economy, but it has the long-term problem that it may produce two different types of worker who – never having experienced each other's problems at first hand – may fail to communicate as well as they should.

WATER DIVINING

No discussion of groundwater exploration would be complete without a reference to **water divining** (also called **dowsing** or **water-witching**). This is the method by which some people claim to be able to locate groundwater by walking over the surface of the ground until they observe a response with a forked stick, bent rods, a pendulum or some other apparatus which is usually held in front of them with both hands.

It is difficult to assess the technique objectively; there is at present no scientific explanation as to why it should work. Moreover, whenever the technique has been tested impartially, the success of the proponents has been no more than would be expected from chance. A dowser can walk across the English Chalk and predict that water will be found below a certain spot; a hydrogeologist knows that a well drilled almost anywhere on the Chalk will encounter some water. However, the subject cannot simply be dismissed. Many people (including some scientists and engineers) can locate buried pipes with the aid of rods or twigs. One theory is that muscles in the body react to some electromagnetic effect caused by the presence of the metal or of the flowing water in the pipe; the rods amplify these minor 'twitches' so that the searcher is aware of them. Another theory is that after long experience in an area, some diviners know intuitively the likely settings in which groundwater will be found near the surface, and subconsciously cause the reaction.

There may or may not be something in these explanations. Even if the electromagnetic theory works for pipes, there is no reason why it should detect the slow, diffuse movement of groundwater in an aquifer.

Perhaps the most telling argument against the use of divining however is that simply locating the presence of groundwater is but a small part of the task of hydrogeology. The hydrogeologist must consider the long-term effects of abstraction, whether it will affect other wells or springs, whether there will be changes in quality with time, and so on. It is difficult to see how the diviner's twig can deal with all these problems.

SELECTED REFERENCES

Anderson, M.P. and W.W. Woessner 1992. *Applied groundwater modeling.* San Diego, California: Academic Press.

Freeze, R.A. and J.A. Cherry 1979. *Groundwater.* Englewood Cliffs, NJ: Prentice-Hall. (See especially Ch. 8.)

Griffiths, D.H. and R.F. King 1981. *Applied geophysics for engineers and geologists.* Oxford: Pergamon.

Keys, W.S. and L.M. MacCary 1971. Application of borehole geophysics to water-resources investigations. *Techniques of water-resources investigations of the US Geological Survey*, Book 2, Chapter E-1. Washington, DC: US Govt Printing Office.

Kruseman, G.P. and N.A. De Ridder 1983. *Analysis and evaluation of pumping test data*, 3rd edn. Bull. 11. Wageningen, Netherlands: International Institute for Land Reclamation and Development.

Price, M. and A.T. Williams 1993. A pumped double-packer system for use in aquifer evaluation and groundwater sampling. *Proceedings of the Institution of Civil Engineers, Water, Maritime and Energy Journal* **101,** 85–92.

Price, M., B. Morris and A. Robertson 1982. A study of intergranular and fissure permeability in Chalk and Permian aquifers, using double-packer injection testing. *Journal of Hydrology* **54,** 401–23.

Tate, T.K., A.S. Robertson and D.A. Gray 1970. The hydrogeological investigation of fissure flow by borehole logging techniques. *Quarterly Journal of Engineering Geology* **2,** 196–215.

Theis, C.V. 1935. The relation between the lowering of the piezometric surface and the rate and duration of discharge of a well using ground-water storage. *Transactions of the American Geophysical Union* **16** (Part II), 519–24.

Todd, D.K. 1980. *Groundwater hydrology*, 2nd edn. New York: Wiley.

Water quality

A SANITARY TALE

In 1854 a cholera epidemic struck Soho, in Central London, causing an estimated five hundred deaths in ten days. Unfortunately, such events were not uncommon around that time; between 1848 and 1850, more than 50 000 people died in a cholera epidemic in Britain.

Central London in the middle of the 19th century was an insalubrious place, with many people supplied with water from the River Thames which also received their untreated sewage. Overcrowding was appalling, and it was commonly believed that diseases like cholera were spread by 'noxious vapours'. A contemporary book went so far as to publish a 'cholera map', which correlated the incidence of cholera with elevation of the land; the incidence of cholera was greatest on the low-lying areas near the river, which were subject to damp and mists. Therefore, went the argument, cholera was clearly caused by the 'mists' or 'vapours'.

One person who had doubts about the 'vapours' theory was an anaesthetist called Dr John Snow. He had pioneered the use of chloroform in place of ether in anaesthesia, and had achieved recognition in his profession by administering chloroform to Queen Victoria at the birth of two of her children. Snow noticed that the low-lying areas in which cholera was most prevalent were all served by two water companies which drew water from the Thames; the higher, healthier areas were supplied from other sources. He reasoned that it could be water, not 'mists', which carried cholera, and the epidemic of 1854 was to give him the proof he needed.

Faced with the spread of the disease Snow did something which today might seem obvious, but at that time was almost revolutionary – he plotted each case of cholera on a map. The points clustered around a hand-pump in Broad Street in Soho, which drew water from a well in the terrace gravels of the Thames, and Snow was able to verify that almost all of the people affected by the disease had drunk water drawn from the well. Two cases stood out – these were at a house in Hampstead, 6 km away. Snow found that the house was occupied by a widow who had moved there from Soho.

She found that the Hampstead water was not to her taste, and so sent a servant each day to collect the familiar water from Broad Street – a small indulgence which cost her own life and that of her niece.

For Snow, his map and discoveries about the well were the final pieces of evidence. He went to a meeting of the Board of Guardians of St James' Parish who, on 7 September 1854, were discussing the epidemic and seeking some way to combat it. What Snow actually said is not recorded, but his advice has passed into legend as 'Take the handle off the Broad Street pump'. This advice was followed, the handle being removed on the following day. By Snow's own account, published the following year, the epidemic had been declining anyway – by then, the majority of the local people had either died or fled – so it is difficult to say categorically that his discovery halted the epidemic. What it did do was to make the authorities aware of the way in which cholera was transmitted, years before cholera bacteria were identified, and in that respect Snow's work probably saved thousands of lives.

The Broad Street pump has long vanished, and Broad Street is now Broadwick Street. At the junction of Broadwick Street and Lexington Street a public house has been renamed the 'John Snow'. It contains a copy of Snow's map and other information connected with the epidemic. Outside, a pink granite kerb stone is said to mark the spot where the pump stood, and a small plaque on the wall records its significance. Much of the surrounding area has been 'redeveloped', and little else is left to provide a link with its place in history, but across the street are buildings that would have been dwelling houses at the time of the epidemic. No doubt some of the victims died inside them.

When Snow was carrying out his investigation in Soho, he was doing more than locate the source of a cholera epidemic. He was beginning the science of **epidemiology** – the study of the way that diseases like cholera are spread. He was also carrying out one of the earliest recorded investigations into a case of groundwater pollution, for it is clear that the water being drawn up by the Broad Street pump had become polluted with human excrement.

We shall look more at pollution of groundwater in Chapter 13. But before we can think about polluted water, we have to think about the natural quality of that water so that we have a baseline to refer to.

GROUNDWATER CHEMISTRY

The study of the chemistry of groundwater is a vast topic that ranges from the routine (but essential) work of the water-supply chemist, who must ensure that the water supplied by his or her undertaking meets the requirements of consumers and statute, to the esoteric studies of the geochemist concerned with diagenetic changes and the slow interaction between groundwater and its host rock.

We saw in Chapter 4 that water is an exceptionally good solvent – it is sometimes called the 'universal solvent' because it will dissolve almost any substance, given sufficient time, even if only in small amounts. To understand some of the reasons for this, we need to look at the atomic structure of water.

Atoms consist of a nucleus containing positively charged protons and (except for hydrogen) other particles called neutrons which have no electrical charge. Around the nucleus, negatively charged electrons move in orbits, rather like planets around the Sun. The charges on electrons and protons are equal, and for any element the number of protons and electrons in an atom is equal and is called the **atomic number** of that element; thus a normal atom has no net electrical charge.

Many atoms find it easy to gain or lose electrons. This upsets the electrical neutrality; they acquire net electrical charges and are called **ions**. Those atoms with net positive charges (those that have lost one or more electrons) are called **cations** and those with net negative charges (if they have gained electrons) are called **anions**. Positive and negative ions can combine to form molecules.

The simplest atom, that of the element hydrogen, consists of a single electron orbiting a nucleus that consists of just one proton. It is relatively easy for the electron to leave or **dissociate** from the proton, which is left as a cation, and to join with another atom to form an anion.

In the water molecule, two hydrogen atoms have combined with one oxygen atom. The molecule is electrically neutral in total, but because of the way the two hydrogen ions are bound to the oxygen ion, one end of the water molecule has a positive charge and the other end has a negative charge; the molecule is said to be an electrical **dipole**.

Molecules of many other compounds, including many rock-forming minerals, are similarly held together by electrical charges. When water flows slowly over the surfaces of these minerals, the charges on the water molecules can be strong enough to overcome the bonds holding the molecules of the mineral together; it is this effect that makes water such a good solvent. The mineral dissociates into ions which enter the water in solution.

The amount of solid material in solution can be expressed in terms of the percentage that it contributes to the total weight of the solution; the solids in sea water, for example, account for about 3.5% of the total weight. In the water in most rivers, lakes and aquifers, the concentrations of dissolved solids are much lower than in sea water, and it would be inconvenient to use percentages. Instead, the weight of solids in each litre of solution is expressed in milligrams, so the unit of concentration is the milligram per litre or mg/l. A milligram is 0.001 gram, and a litre of pure water at most ground temperatures weighs almost exactly 1000 grams; provided that the concentration of solids in the solution is not so high as to affect the density greatly, one milligram per litre is therefore equivalent to one part per million (ppm) by weight.

The total amount of solid material dissolved in a sample of water can be determined by evaporating a known volume of water to dryness, and then further drying the residue in an oven, usually at a temperature of 180 °C. This tells us how much material is present in solution, but not what the dissolved substances are. When water samples are analysed, the results are usually expressed in terms of the ionic concentrations. For example, when 1 g of sodium chloride (NaCl) is dissolved in pure water and made up to 1 litre of solution, the result is expressed as

Constituent	Concentration (mg/l)
Na^+	393.4
Cl^-	606.6

The plus sign after the Na symbol shows that dissolved sodium occurs as cations, and the minus sign after the Cl symbol shows that dissolved chloride occurs as anions.

Some substances dissolve more readily in pure water than do others. Sodium chloride (common salt) is highly soluble; silica is only slightly soluble.

Although water molecules are chemically relatively stable, some dissociate into hydrogen (H^+) and hydroxyl (OH^-) ions. The dissociation can be represented

$$H_2O = HOH = H^+ + OH^-$$

Thus even pure water contains some ions, although in very small concentrations. Some substances, such as sodium chloride, dissolve in water without causing any change to the concentrations of hydrogen and hydroxyl ions, but others have a marked effect. A substance that liberates hydrogen ions in solution in water is called an **acid**, and the resulting solution **acidic;** a substance that liberates hydroxyl ions in solution in water is called a **base**, and the resulting solution is said to be **alkaline**. A solution in which the numbers of hydrogen and hydroxyl ions are equal is said to be **neutral**. The concentrations of these ions in natural waters are generally too low to express in the usual terms of mg/l, so a special scale is used. This is the **pH** scale (the H denoting hydrogen), which is a measure of the hydrogen-ion activity of the solution. (It is defined as the logarithm of the reciprocal of the hydrogen-ion activity measured in moles per litre.)

The pH of a neutral solution varies with temperature: at 25 °C it is 7. Solutions with a pH less than 7 are acid, those with a pH greater than 7 are alkaline. Because the scale is logarithmic, a change of one point on the pH scale implies a tenfold increase or decrease in acidity.

Pure water thus has a pH of 7, but water does not stay pure for very long. Distilled water, for example, left in contact with the atmosphere, is found to have a pH of between 5.5 and 6. This is because carbon dioxide from the atmosphere dissolves in the water and leads to the dissociation of water molecules to liberate hydrogen ions. This reversible reaction is the first stage of the process by which rain water can dissolve calcium carbonate as described briefly in Chapter 4:

$$H_2O + CO_2 = H^+ + HCO_3^-$$

The carbon dioxide dissolves to yield hydrogen ions and bicarbonate ions; the solution is called carbonic acid.

Weak acid solutions have the ability to dissolve some substances that are relatively insoluble in pure water. This is what happens, for example, when limestone (calcium carbonate) is dissolved in carbonic acid, H_2CO_3 (Chapter 4).

Calcium carbonate does not readily dissociate into calcium and carbonate ions, so it dissolves slowly and with difficulty in pure water. If, however, a hydrogen ion approaches the interface between the calcium carbonate and the water it can partially neutralize the negative charge on the carbonate anion:

$$CaCO_3 = Ca^{2+} + CO_3^{2-}$$

and
$$H^+ + Ca^{2+} + CO_3^{2-} = Ca^{2+} + HCO_3^-$$

Unless a hydroxyl ion is also close by, the solution will be electrically unbalanced and another carbonate ion must dissociate to restore the balance. In pure water (pH 7), the number of hydrogen and hydroxyl ions will be the same, and so it is likely that there will be a hydroxyl ion present to restore the balance and inhibit further dissociation and dissolution. In acidic water, by definition, the excess of hydrogen ions means that dissolution is likely. Notice that the carbonic acid contains hydrogen ions and bicarbonate HCO_3^- ions: the process of dissolution yields calcium and more bicarbonate ions, the latter formed by the hydrogen ions from the carbonic acid combining with carbonate groups.

Carbonic acid is relatively weak. Another natural process can produce much stronger acid. Many rocks, especially mudrocks such as clays and shales, contain small amounts of the iron sulphide (FeS_2) minerals, pyrite (fool's gold) and marcasite. In the presence of oxygen these can be oxidized to form iron sulphate and sulphuric acid. Although the reaction can take place inorganically, sulphur-oxidizing bacteria are generally thought to play a major part. The reaction takes the form

$$2FeS_2 + 2H_2O + 7O_2 = 2FeSO_4 + 2H_2SO_4$$

Reactions of this type can occur whenever these sulphide minerals are exposed to air. The drainage of mine workings, for example, can lead to the entry of air and the generation of acidic waters. When discharged at the surface, or when the mines are abandoned and refill with water so that the water discharges naturally, these acidic waters can cause severe problems to the ecology of streams.

Table 11.1 Typical pH values

	pH
Battery acid	1
Lemon juice	2.2
Vinegar	3
'Acid' rain	2–5
'Natural' rain	5.6
Pure water	7
Sea water	8.3

With strong acids such as hydrochloric acid or the sulphuric acid generated as described above, a second process of solution may come into effect. A chemical reaction may take place between the acid and the solid in which the products include water and a soluble salt.

In London, acid waters have attacked the linings of underground railway tunnels; these waters are believed to have been generated by air, forced through the tunnel linings by the passage of trains, reacting with the pyrite in the adjacent formation. Corrosion of the linings created more spaces for air to leave the tunnels, making the situation worse.

'Natural' rain water, like distilled water in contact with the atmosphere, has a pH of between 5.5 and 6 because of its content of dissolved carbon dioxide (Table 11.1). In many areas rainfall has become more acidic because it also dissolves gases such as sulphur dioxide and various oxides of nitrogen. These are produced by industry and the burning of hydrocarbons in vehicles or to generate electricity. In and downwind of many cities and industrial areas the pH of rainfall may be less than 4, and values as low as 2 have been reported; in parts of Europe and North America this **acid rain** is viewed as a major environmental problem.

Despite the acid content of rainfall, and the fact that water usually dissolves more carbon dioxide as it infiltrates the soil, most groundwaters have a pH between 7 and 8. They are thus neutral or slightly alkaline. The widespread occurrence of calcium carbonate in many sedimentary aquifers, either forming the rock framework in limestones, or present as cement in many sandstones, is largely responsible for this. It also means that many groundwaters are relatively well buffered – they tend to resist changes in pH. This is fortunate because many metals are soluble in acid waters, and many metals are toxic to humans, animals and some plants.

ORIGINS OF GROUNDWATER

Most groundwater abstracted for domestic, industrial or agricultural use is **meteoric** water, i.e. groundwater derived from rainfall and infiltration within the normal hydrological cycle. The word meteoric comes from the same root as 'meteorology' and implies recent contact with the atmosphere. As we shall see later, the chemistry of meteoric groundwater changes during its passage through rocks, the changes depending on such factors as the minerals with which it comes into contact, the temperature and pressure conditions, and the time available for water and minerals to react. The modification of meteoric groundwater in its passage into and through the ground is one of the evolutionary sequences of groundwater chemistry, and it occurs in many aquifers.

The generally saline waters encountered at great depths in sedimentary rocks were at one time all thought to have originated as sea water trapped in marine sediments at the time of their deposition, and were referred to as **connate** waters. It is now accepted that meteoric waters may eventually become saline. It is also accepted that most of the original sea water has been modified and has moved from its original place of entrapment; this evolution of marine waters is the second sequence of change of groundwater chemistry. There is now some debate as to whether the term 'connate' should be used to refer to entrapped marine water in its original sediments, to marine water that may have migrated, or to any saline groundwater; or whether indeed it should be used at all. Most workers would agree that the term 'connate' implies that the water has been removed from atmospheric circulation for a significant (in geological terms) length of time; I shall use it to denote groundwater derived mainly or entirely from entrapped sea water.

There is a third possible origin for groundwater. **Juvenile** water is the name given to the water believed to be derived from igneous processes within the Earth, and which can contribute unusual constituents to the meteoric groundwater which it joins. According to strict definition juvenile water has never previously taken part in the hydrological cycle; one theory is that all the Earth's water originated at one time as juvenile water. Others point out that juvenile water is indistinguishable from meteoric water that has penetrated to great depths and become intimately associated with igneous processes; it is therefore possible that much or all of this supposed 'juvenile' water is really meteoric in origin. It is also possible that much of the water released during igneous events originates as oceanic or connate water trapped within rocks carried beneath the Earth's crust, as crustal plates collide and sink beneath each other at subduction zones. This still leaves us with the question of how the Earth's water did originate, but that is a topic for cosmologists as much as for hydrogeologists.

CHEMICAL DEVELOPMENT OF METEORIC WATER

Meteoric water, by definition, was once precipitation. Although precipitation is nature's form of distilled water, it typically contains between 10 and 20 mg/l of dissolved material. Near coastlines the concentration of sodium chloride is increased, and downwind of industrial areas sulphur and nitrogen compounds are more in evidence.

When the precipitation infiltrates the soil, the most important natural change is the dissolution of carbon dioxide from the soil atmosphere. The soil organisms also consume much of the oxygen that was dissolved in the precipitation in its passage through the atmosphere.

In temperate and humid climates, where recharge is a regular process, water is usually moving relatively rapidly through the outcrop area of an aquifer, so its contact time with the rocks may be limited. Any highly soluble materials, such as sodium chloride, will have been flushed from the system long ago, and there will often be insufficient time for poorly soluble minerals to be taken into solution in significant amounts. As a result, groundwater in the outcrop areas of aquifers in countries like Britain tends to be low in dissolved solids unless the aquifer is a limestone or a sandstone with calcium-carbonate cement, in which case the groundwater will contain calcium and bicarbonate as the dominant ions.

Sulphate and nitrate will also be present in solution in small quantities. Nitrates are usually derived from the soil, where they may originate from the fixation of atmospheric nitrogen by leguminous plants or from the oxidation of organic materials by bacteria. Sulphate ions are commonly produced in the outcrop area by the oxidation of metallic sulphides that are present (at least in small quantities) in many rocks. The common sulphate minerals – gypsum and anhydrite – are readily soluble, and like sodium chloride will usually have been flushed from the outcrop area already.

If the aquifer dips below a confining bed, the conditions prevalent at the outcrop may continue for some way below the impermeable cover. Dissolution is still the dominant process, and in aquifers containing calcium carbonate the amounts of calcium and bicarbonate in solution typically rise to their highest levels in this part of the aquifer.

With increasing distance from the outcrop, the dominant process changes from dissolution to ion exchange. Most aquifers contain some clay minerals; the small size of clay particles means that, although the clay may be present in only small total quantities, it presents a relatively large surface area to the percolating groundwater. Ions adsorbed on the clay surfaces tend to exchange with ions in solution; this principally involves positive ions, the trend being for the cations in the water to come into equilibrium with those on the clay particles in the aquifer. A major effect is that calcium and magnesium ions in solution are replaced by sodium ions, which are frequently concentrated on clay surfaces when rocks are deposited. The removal of calcium ions from the water can lead to further solution of calcium carbonate, but the end product tends to be a sodium-bicarbonate water. Since it is calcium and magnesium ions that cause hardness, ion exchange leads to a softer water.

At greater distances from the aquifer outcrop there is typically less natural movement of water because there is generally no outlet for the water

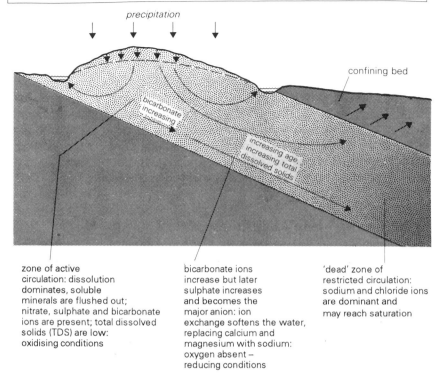

precipitation

confining bed

bicarbonate increasing

increasing age, increasing total dissolved solids

zone of active circulation: dissolution dominates, soluble minerals are flushed out; nitrate, sulphate and bicarbonate ions are present; total dissolved solids (TDS) are low: oxidising conditions

bicarbonate ions increase but later sulphate increases and becomes the major anion: ion exchange softens the water, replacing calcium and magnesium with sodium: oxygen absent – reducing conditions

'dead' zone of restricted circulation: sodium and chloride ions are dominant and may reach saturation

Figure 11.1 The general chemical evolution of meteoric groundwater.

as the strata dip downwards, except by leakage through the confining beds. The slower movement means that there is more opportunity for less-soluble minerals to be dissolved. The reduced throughflow also means that soluble minerals such as gypsum may not have been flushed from the aquifer; with even greater distance from the outcrop and greater depth of burial, this will even be true of sodium chloride. Thus there is a change from bicarbonate waters to sulphate waters and finally to chloride waters.

To summarize (Figure 11.1), a complete typical sequence of the evolution of meteoric groundwater starts with water in which the main anions are bicarbonates. As the water moves deeper, sulphate ions increase in importance and become dominant; finally, if the system is deep enough, chloride becomes dominant. The amount of material in solution also increases with time and depth. The sequence in which the various cations become important is complicated by the effects of ion exchange. A broad generalization, however, is that as the groundwater becomes older and deeper the dominant cations change from calcium and magnesium to sodium.

It must be emphasized that the so-called 'typical' sequence is subject to many variations. For one thing, the water may not stay in the aquifer long enough or move deep enough to reach the 'typical' end product of a sodium-chloride brine. Conversely, if water exceptionally infiltrates an arid region where evaporite deposits are still present, it may reach the sodium-chloride stage almost immediately. The general idea of the sequence is nonetheless a useful one. It is sometimes called the **Chebotarev sequence**, after the scientist who first proposed it.

Another significant change that usually occurs with increasing distance from outcrop (and therefore usually with increasing depth) is the change from oxidizing to reducing conditions. As noted previously, groundwater at the outcrop will generally contain oxygen (derived from the atmosphere) and nitrate and sulphate from the soil and outcrop. As the water containing dissolved oxygen, nitrate and sulphate moves away from the outcrop, oxygen is used up in oxidizing organic matter and other material such as ferrous iron. Dissolved oxygen is used first, and then nitrate and finally sulphate are reduced as their oxygen is used up. As with the oxidation of sulphides, micro-organisms play a role in these reduction reactions. Like the evolution of the major ion chemistry, the time and distance taken for these changes will vary from one aquifer and place to another.

Desert-lain sandstones (such as many of the British Permo-Triassic sandstones) contain very little organic material, and as they were deposited under generally oxidizing conditions they have little ability to remove oxygen from groundwater. In these aquifers it is therefore common to find oxidizing groundwater at considerable distances from the outcrop area.

In contrast, other aquifers may contain large amounts of organic material, which rapidly depletes the oxidizing capacity of the groundwater. In the Lincolnshire Limestone (of Jurassic age) in eastern England, for example, there is a fairly abrupt change from oxidizing to reducing conditions about 12 km from the beginning of confined conditions. In this aquifer and in many others, the presence of reducing conditions results, among other indicators, in the presence of hydrogen-sulphide gas.

GROUNDWATER QUALITY IN ARID AREAS

The above discussion relates to temperate or humid conditions, where there is an annual surplus of precipitation over evapotranspiration. Aquifer outcrops will usually have been flushed to a depth of several hundred metres or the full thickness of the aquifer by water moving from recharge areas to discharge areas, and the water will generally have a low dissolved-solids content.

In arid and semi-arid areas (Figure 11.2) there may be significant differences. Because evaporation or evapotranspiration generally exceeds rainfall for most or all of the year, the aquifer outcrop may be a zone of concentration rather than of dissolution. The dissolved solids content of the rainfall may be small, but over thousands of years, as water is evaporated and the solids remain in the ground, the salinity of any recharge water increases markedly. In addition, any soluble minerals in the aquifer may not have been leached out, and soluble weathering products may not be removed as

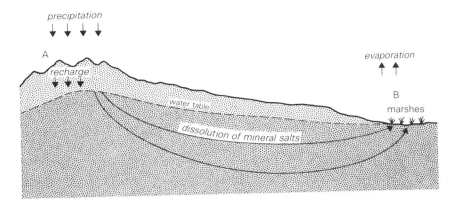

Figure 11.2 Groundwater flow in an arid region. In this example, recharge takes place only in the mountains, and discharge occurs in low-lying areas giving rise to salt marshes.

quickly as they form; these too may increase the dissolved solids content of the groundwater.

If the water table is sufficiently close to the ground surface for groundwater to be lost either by capillary action and direct evaporation or by transpiration, there is likely to be a concentration of salts within the soil and subsoil, as water is evaporated or transpired and the solutes are left behind. In this way a crust or 'hardpan' can form at or near the surface.

In such conditions irrigation, far from helping, may aggravate the problem. If groundwater is used for irrigation in arid areas, then the high rates of evaporation mean that soluble material in the groundwater will be deposited in the soil. Any occasional natural recharge or excess of irrigation water is liable to dissolve these minerals and carry them down to the water table, thereby increasing the salinity of the groundwater.

In the Indus Valley of Pakistan the groundwater in the alluvial aquifer is naturally saline. A network of canals, the largest bigger than the Suez Canal, carries water from the Indus to irrigate the adjacent land (Figure 11.3a). The canals divide repeatedly into smaller branches which distribute the water to the fields. The system used is flood irrigation, which in essence means that the fields are periodically inundated. In ensuring that water reaches the higher parts of each irrigation area, it often happens that the lower areas receive too much water. The surplus infiltrates and mixes with the native groundwater. Over the decades this has resulted in the water table in some places rising very close to the surface, causing the 'twin evils' of waterlogging and salinization of the soil. As a result, much land has been lost to agriculture.

Salinization destroys agricultural land partly because the plants cannot tolerate saline water, and partly because the high concentrations of sodium affect clays, causing the soil to lose its 'crumb' structure and reducing its

Figure 11.3 Problems caused by irrigation in the Indus Valley. (a) The main canals that distribute water from the River Indus in Pakistan are as large as British rivers. They are unlined and fresh water leaks from them into the underlying alluvial aquifer. This water causes the water table to rise and leads to waterlogging and salinization of the soil. (b) Locally the infiltrating fresh water is exploited by sinking wells fitted with hand-pumps near the canals. These hand-pumps provide supplies for villages.

permeability. This in turn may increase the problem by inhibiting drainage further and perhaps leading to the presence of perched saline water. To help restore some of the land affected in this way in the Indus Valley there is now a programme to lower the water table, using a combination of wells and drains to pump the saline water and convey it to the coast.

Studies for these projects have revealed that so much water leaks from the canals that, along many of them, there is frequently a layer of fresh water lying above the denser saline water. Local people often exploit this fresh groundwater by sinking shallow wells and installing hand-pumps to provide village supplies (Figure 11.3b).

Waterlogging and salinization of the soil, with consequent loss of production, has also been a problem in other areas. When land that was previously well-drained and fertile becomes waterlogged or salinized, there has clearly been some change to the water balance of the area. This can happen not only because too much water is applied, as in the Indus Valley, but when less water is removed. A common cause is when perennial plants, especially trees, are cut down and replaced by seasonal crops. The reduction in use of water can lead to a significant rise in the water table.

This has been a problem in Western Australia, where clearance of eucalyptus trees has been a major factor. The roots of eucalyptus reach the water table, so it is particularly effective at maintaining low groundwater levels. In the Murray Basin, in south-eastern Australia, a combination of irrigation and land clearance has led to salinization destroying the agricultural capacity of large areas of land. In both the Murray Basin and the Indus Valley, the solutes in the groundwater are believed to have originated in precipitation, the concentration having increased as water has been evaporated or transpired.

The limited recharge in arid and semi-arid regions often means that groundwater has either migrated from outside the area or infiltrated a long time ago (Chapter 8). Either alternative implies long residence times and possibly incomplete flushing of soluble minerals from the aquifer. Sodium-chloride waters are therefore relatively common. A typical situation might be that of Figure 11.2. Here, recharge on the mountains at A enters and percolates through the aquifer, dissolving soluble materials in its passage, until it emerges (perhaps thousands of years later) in seepages at B. Evaporation is too great to allow the formation of a perennial stream at B, so the discharge area is likely to be stagnant marsh or ponds. Sodium chloride is likely to be the dominant constituent, so salt marshes (like the inland **sabkhas** of North Africa) are a common feature. For the same reason, many of the Saharan oases yield brackish or saline water. Evaporation may lead to the formation of saltpans or beds of evaporite minerals which in the course of time may become part of an evaporite sequence. It is the dissolution of minerals of this type, formed in the geological past, which is thought to be responsible for the high salinities of many present-day deep groundwaters.

CONNATE AND SALINE WATERS

Most meteoric groundwaters abstracted for water supply have total dissolved solids (TDS) concentrations of a few hundred milligrams/litre or less. Anything higher would have an unacceptable taste. At depth in some formations the TDS concentration may rise to the level of saturation, which for sodium chloride at typical temperatures is about 250 000 mg/l. The higher concentrations (above a few thousand mg/l) are nearly always dominated by sodium chloride. Since such waters are not abstracted for supply purposes the formations that contain them are not aquifers in the normal sense. Knowledge of the waters is usually obtained from mine drainage, or from oil and gas exploration and production programmes.

Some of these brines may be formed by meteoric water dissolving the salt deposits laid down in earlier geological periods, but until recently many were thought to have originated from sea water trapped in sediments when they were deposited (connate waters). However, since the TDS composition of sea water is at present only about 34 000 mg/l, and it has probably not been significantly higher in the past, some mechanism is needed to explain how the connate water has become concentrated. It has been suggested that the concentration could come about by a natural form of reverse osmosis.

If two solutions of similar composition but different strengths are separated by a **semi-permeable membrane** (a membrane that permits the passage of a solvent but not of the dissolved substances) there is a natural tendency for solvent to pass through the membrane from the weaker to the stronger solution. Thus, in the case of two sodium-chloride solutions, water would flow through the membrane to dilute the stronger solution; the weaker solution would become more concentrated and the process would end when the solutions were of the same strength. This process is called **osmosis**. The fact that water flows naturally through the membrane implies that an energy difference must exist as a result of the different chemical concentrations.

Conversely, if the head of the stronger solution is artificially increased (by putting a high pressure on it, for example), solvent can be forced to flow through the membrane from the more concentrated to the less concentrated solution. This process is termed **reverse osmosis** and is used to produce fresh water from sea water in desalination plants. A semi-permeable membrane can thus be thought of as a 'sieve' that permits the passage of some ions but not others.

As marine sediments are consolidated by burial beneath other deposits, the sea water they contain would normally be 'squeezed' from them and would generally migrate in an upward direction. However, clays and shales may act as semi-permeable membranes, allowing the water to pass but retaining some of the ions present in the solution. The sieve mechanism actually seems to be electrical in nature: negative charges on the surface of the clay particles repel (and hence prevent the passage of) the negatively charged ions (e.g. the chloride ions). The positively charged ions must also be held back, otherwise the solution would be ionically unbalanced; hence most of the water passes through the clays and the dissolved solids are retained in the pores of their parent sediment. The remaining pore water thus becomes progressively more saline. The mechanism has been reproduced in laboratory experiments and seems attractive, although recent work has suggested that most saline groundwaters originate by the dissolution of evaporite minerals.

Presumably, solutions derived from meteoric waters could also become concentrated by this **membrane-filtration** effect, given appropriate conditions.

ISOTOPES AND TRACERS

It should be apparent by now that the length of time that water spends in contact with the rock, from infiltration to discharge, can have a considerable effect on its quality. Knowledge of this residence time can also be important in assessing present-day recharge and pollution risks; did the water that we pump from a well today enter the aquifer last month, last year or 10 000 years ago? If the last, can we be sure that recharge is occurring now to replenish that water, or are we 'mining' a finite resource? In many places we have endangered the quality of groundwater by dumping poisonous waste materials where leachate from the wastes could reach an aquifer. How long will it take for the toxins to reach an important aquifer, or to pass through a natural flow system to a river?

Groundwater can move through aquifers with speeds ranging from many metres a day to less than a metre a year; this range is possible even within the same aquifer. In confining beds the speeds are, of course, usually slower. How can the residence time of the water be quantified? It would obviously be useful to be able to label a droplet of water with date, time and place, introduce it into an aquifer and await its discharge, much as an ornithologist charts the movements of a ringed bird.

Labels cannot be fixed to water droplets, but for many years cave explorers have used dyes and spores to 'label' the waters of streams that flow through caverns for parts of their courses, and so determine their unseen routes. In principle, the idea can be applied to the study of groundwater; the technique is known as **tracing**, and the substance added to the water to distinguish it is a **tracer**. The substance used for the tracer must move as part of the water, at the same rate, and not become filtered or separated from it; it must not be present naturally in the water (or at least not in the concentrations at which it is being used as a tracer); it must be readily detectable, easy to handle, non-toxic and, ideally, cheap. The fluorescent dyes popular with cave explorers tend to be adsorbed on clays, and the lycopodium spores that the speleologists also use are too dense and too large to be carried by laminar flow through the pores of an aquifer.

The ideal tracer, in short, does not exist, but many substances have been used to trace the movement of groundwater and to indicate its natural flow speeds over distances up to a few tens of metres or, in the case of flow through large fissures, a few kilometres. Successful tracers have included fluorescent dyes and tiny micro-organisms called bacteriophages (p. 196). Other micro-organisms, including bacteria and yeasts, have also been used. Among the more useful fluorescent dyes are a variety of fluorescein called uranine, a dye called Rhodamine WT, specifically developed for water tracing, and some of the optical brighteners used in washing detergents. The optical brighteners are the 'miracle ingredient' added to most washing powders since the 1960s to make 'whites' look whiter; they do this by fluorescing, absorbing ultra-violet light and emitting the energy as visible light. As tracers they have the advantage that, because they are produced in bulk for industry, they are relatively cheap; they are also non-toxic. They are adsorbed by textiles which then fluoresce in ultraviolet light, providing a simple method of detection. Like the other fluorescent dyes, they can also be detected at very low concentrations in water samples using an instrument called a spectrofluorometer; in ideal conditions some dyes can be detected in water at concentrations as low as 1 part in 10 000 million.

Over distances greater than about a hundred metres (or, exceptionally, a few kilometres where flow in large fissures is involved) new problems arise. The tracer becomes so dispersed among the bulk of the groundwater that it is likely to be undetectable, and the time taken becomes too great – in some cases many times greater than a human lifetime, so that water pumped from a well today would have to have been 'labelled' perhaps thousands of years ago for us to assess its residence time in the aquifer.

Fortunately, nature comes to our aid by labelling water for us; the information that we need is affixed to every drop of water that enters an aquifer – if we can only read the label! The labels that nature provides are 'isotopes'. All the atoms of a given element have the same atomic **number** – their nuclei have the same number of **protons**: however, some have different **atomic mass**, because they have more or fewer **neutrons**. These different forms of the same element are called **isotopes**.

Most elements are mixtures of two or more isotopes. Some isotopes are stable; others (**radio-isotopes**) change by radioactive decay into isotopes of other elements.

The rate of this decay varies from one isotope to another, but for each isotope it is more or less constant and is known fairly accurately. This means that if a radio-isotope of a particular element is present in recharge water in a known ratio to the stable isotope of that element, then during its passage through the aquifer the proportion of the radio-isotope will decrease as that isotope decays. If no more of it is added to the water, it is possible to determine how long it has taken the recharge water to reach any point in the aquifer by measuring the new ratio of radio-isotope to stable isotope at that point.

For example, a radio-isotope of carbon, ^{14}C (carbon-14) is formed continuously by the effects of cosmic rays on nitrogen in the upper atmosphere. Until relatively recent fossil-fuel burning, and the testing of thermonuclear weapons which began in 1952, the proportion of ^{14}C to stable carbon present in the atmosphere as carbon dioxide was believed to have been more or less constant for tens of thousands of years, the decay of ^{14}C being balanced by its production. Carbon in rain and in soil water (as dissolved carbon dioxide) is present in the same isotopic proportions. Once this water enters an aquifer it is no longer in contact with the atmosphere. The ^{14}C decays, so that the proportion of ^{14}C to stable carbon decreases. The magnitude of this ratio in a sample of groundwater is thus an indication of the length of time since the water left the atmosphere.

Unfortunately, not all the carbon in groundwater comes from the atmosphere or the soil. As we have seen, some can come from dissolution of calcium carbonate, which has probably been part of the aquifer for so long that it contains only stable carbon. Corrections can be made for this, but for these and other reasons isotopic dating (also called **radiometric dating**) of groundwater is never exact and is occasionally completely unreliable. It should be regarded as giving only an approximate guide to the time that water has been in an aquifer.

Carbon-14 has a half-life of about 5600 years, which means that in any sample containing ^{14}C, half of the ^{14}C nuclei will disintegrate in 5600 years. Radiometric dating using ^{14}C is therefore useful for waters that have been underground for some thousands of years, up to a maximum of about 30 000 years. The method has been used to indicate that much of the groundwater present beneath the Sahara desert infiltrated thousands of years ago (p. 121).

Another isotope of use in groundwater studies is **tritium**, a heavy isotope (^3H) of hydrogen. (Normal hydrogen is ^1H.) Tritium, like carbon-14, is produced naturally in the upper atmosphere by the action of cosmic rays on nitrogen. The concentration of tritium is measured in tritium units (TU), one tritium unit being a concentration of one tritium atom in every 10^{18} hydrogen atoms. Under natural conditions the tritium content of precipitation is less than 10 TU. However, like carbon-14, tritium is also formed as a result of the explosion of thermonuclear weapons; between 1952 and 1964 the tritium content of rain was increased considerably, to levels above 2000 TU in parts of the Northern Hemisphere.

- The half-life of tritium is 12.4 years, so the presence of tritium concentrations in excess of 5 TU in groundwater implies that the aquifer has received recharge since thermonuclear tests began in 1952.

The measurement of tritium concentrations requires sensitive, expensive equipment. But for this, tritium has been considered by some workers to be the closest thing to the ideal tracer, since it actually forms part of the water molecule.

One of the conditions laid down for an ideal tracer, such as a radio-isotope, is that it should move at the same rate as the water and be affected in the same way by chemical and physical processes; it should not, for example, diffuse through shales at different rates or be evaporated at different rates, otherwise it may be subject to unquantifiable concentration or dilution. Such changes in concentration are termed **fractionation**. For a radio-isotope such as tritium to be useful, it is necessary that it should not be subject to significant fractionation, and that its concentration should change only as a result of radioactive decay.

There are however other isotopes – stable isotopes – which are useful in hydrological studies precisely because they *are* subject to fractionation and because they are not radioactive and do not change with time. Two isotopes of particular interest to hydrologists are those which can occur in water molecules – a stable isotope of oxygen, ^{18}O (oxygen-18), and a stable isotope of hydrogen, ^{2}H (commonly called deuterium). The proportions of these isotopes to the common stable isotopes, ^{16}O and ^{1}H, can be measured with a mass spectrometer, and expressed relative to their concentration in an arbitrary standard water ('standard mean ocean water' or 'SMOW').

Because of their greater masses, ^{18}O and ^{2}H isotopes are less likely to evaporate and are more likely to condense than are the more usual ^{16}O and ^{1}H isotopes. As water evaporates from the seas and moves inland to begin its journey through the water cycle, many evaporations and condensations may take place. Generally, the further the water moves from the ocean, the lower is its concentration of these heavy isotopes. The fractionation is also temperature-dependent.

Once the precipitation reaches the ground and infiltrates, no further perceptible fractionation occurs unless the water percolates to great depths and reacts with the rock. The ratios of the heavy isotopes to the common isotopes therefore remain unchanged. In the right circumstances, knowledge of the isotopic ratios of groundwater can provide information about the recharge area from which it was derived or – in large basins where waters from several sources become mixed – about the proportions of water from different sources.

When soil water evaporates, the remaining water becomes enriched in ^{2}H and ^{18}O isotopes. The relative enrichment in the two isotopes is different from that which occurs during condensation. Study of isotopic fractionation of soil water is thus a powerful tool for studying soil-water processes, especially in arid regions where evaporation rates are high and fractionation is pronounced.

RESIDENTS, IMMIGRANTS AND VISITORS – THE MYSTERIOUS WORLD OF MICRO-ORGANISMS

When Snow tracked the source of the Soho cholera epidemic to the Broad Street pump (pp. 172–3), he may have suspected that the disease was caused by something in the water, but he could not know what that 'something' was. Cholera is now known to be caused by bacteria which enter the intestinal tract of infected individuals and then multiply rapidly. The bacteria produce a toxin that prevents the body absorbing water from the large intestine, and also upsets the balance of salts, leading to a watery diarrhoea. Unless treatment is at hand, the victims are severely weakened and may die of dehydration. During the illness, the watery excreta of these individuals contain cholera bacteria which, in the absence of adequate sanitation, can be carried into watercourses. If this water is drunk without being boiled or treated other people become infected. The Soho epidemic was caused by untreated sewage entering the shallow groundwater of the terrace gravels.

Other intestinal diseases caused by water-borne bacteria include typhoid, bacillary dysentery and some forms of diarrhoea. Amoebae can cause other forms of dysentery, and two particularly unpleasant forms of diarrhoea are caused by protozoans, *Giardia* and *Cryptospiridium*. Outbreaks of *Giardia* are common in the United States, but rare in Britain. Infectious hepatitis (jaundice) is a virus disease that can be spread by water (although it can be transmitted in other ways) and viruses are responsible for a range of intestinal diseases that cause diarrhoea or vomiting. All of these water-borne diseases are spread in much the same way as cholera, by the so-called faecal–oral route. To prevent the spread of these diseases, water must be tested before it is supplied to the public.

In practice, because it takes a long time to locate and identify disease-producing (**pathogenic**) organisms, a simpler technique of testing is adopted. This involves looking for **indicator organisms** that may or may not be pathogenic in themselves, but which indicate that the water may be contaminated with pathogens. All humans have bacteria of a species called *Escherichia coli*, one of a group known as coliform bacteria, living in their intestines. The presence of coliform bacteria in water implies that the water *may* have been contaminated with human excreta, which *may* have contained pathogenic organisms. In practice therefore the first microbiological test carried out on a water sample is usually for the total or **presumptive coliform count**.

There are various ways of carrying out such a test. A relatively simple one is to introduce a fixed volume of the water to be tested to a plate or filter pad that carries a culture medium that will suit the bacteria being tested for. After the appropriate time (usually 24 or 48 hours), the presence of bacteria in the sample is indicated by the formation of colonies around each original bacterium. The colonies can usually be seen with the naked eye, and the number of bacteria is expressed in terms of **colony-forming units** or **CFUs** in one millilitre of water. Depending on the disease and the general health of the victim, it may take dozens or a million or so bacteria to cause the infection.

Tests for bacteria are usually carried out with cultures at temperatures of 22 °C (room temperature) and 37 °C (human body temperature). A second set of tests will usually be carried out to look specifically for human faecal bacteria – the **faecal coliform count**. There are tests available to identify specific bacteria, including examining them with a microscope.

Where tests have to be undertaken for viruses, essentially the same principle is used. Complications are that viruses reproduce only in living tissue, and are specific as to the type of cells that they attack: in essence, a virus is 'alive' only inside its host. Tests therefore have to be carried out on appropriate living cells. If a plate test is used, the viruses cause patches or **plaques** of dead cells as they reproduce. Each plaque is assumed to result from one initial virus, and the concentration of viruses in water is expressed in terms of the number of **plaque-forming units (PFUs)** found in a litre of water.

A further complication is that viruses are too small to be seen with an optical microscope. A major problem from the health point of view is that viruses cannot be killed by the antibiotics that have proved so effective against many bacterial diseases.

The term **micro-organism** is applied to any organism that we need a microscope to see. It includes bacteria, microscopic fungi, viruses and certain other organisms such as amoebae and some protozoans. **Microbe** is a slightly less formal term. The long strand structure of fungi, and the size of amoebae and protozoans, limit their mobility in the small pore spaces of most aquifers. Since viruses are active only in living tissue, and that tissue is absent from most aquifers, their main interest to the hydrogeologist is in the possible survival and transport of pathogenic varieties in aquifers. Although viruses are really 'alive' only in living tissue, they can continue to survive outside a host in a resting stage of their life cycle, and become active again when they enter a suitable host.

In aquifers, bacteria are the most important group of micro-organisms, but although some species of bacteria can apparently flourish in some aquifers, groundwater is usually free from pathogenic micro-organisms. Pathogens are in essence visitors to the aquifer, which is not their natural habitat; they are just passing through on the way from one human host to another.

Bacteria can come in many shapes and sizes, but generally they are around 0.5 to 5 micrometres (μm) in size, with 0.2 μm taken as the lower limit. This means that the size of the average bacterium is about the same as the size of the interconnections between the pores in a fine- to medium-grained sandstone.

For such a bacterium to travel 1 metre through such a sandstone is rather like you or me having to find our way for 500 km through a maze whose openings are just wide enough for us to squeeze through; the task would be difficult enough if we were well supplied with food. For pathogenic bacteria, deprived of the nutrition they obtain inside the human body, it is generally impossible. Like us, the bacteria will starve and get smaller, which makes it easier for them to move through the pores, and unlike us they are able to detect to some extent the direction in which nutrients are available, and move towards the food source. However, long before they can reach an exit from the aquifer they will usually have died within the pore space. It is worth noting though that many micro-organisms are opportunistic pathogens; their natural home may not be in the body, but given the chance they will settle down there quite happily, frequently to our detriment. The micro-organism that causes tetanus, for example, usually lives in the soil but will infect us through open wounds.

The difficulties faced by 'visiting' micro-organisms trying to survive in aquifers led many people to conclude that aquifers were generally devoid of microbial populations. For many years there seemed to be no direct evidence to refute this conclusion. The soil is known to be well-populated by microbes, but the population drops off rapidly below its base. Indirectly there was the indication that the filtration effect would remove bacteria, and a conviction among most microbiologists that the necessary nutrients were lacking. One of these essentials is carbon (see Box). Conversely, there was the knowledge that various species of bacteria can survive in harsh environments and indications that various processes of sulphate reduction and nitrate reduction taking place in deep aquifers were possibly being controlled by microbes. The sequence in which oxygen, nitrate and sulphate disappear from groundwater migrating down dip, for example, is the sequence in which bacteria would use these compounds for respiration. There was evidence from the oil industry of bacterial activity in some oil fields, and a conviction among some hydrogeologists that bacterial processes were taking place in some aquifers, but it was only in the early 1980s that improved methods of sampling and of testing for non-pathogenic micro-organisms led to conclusive proof that bacteria were indeed present at depth in aquifers, and hydrogeologists and microbiologists began to work together to investigate the subject further.

Many issues remain unresolved, but it is clear that in addition to micro-organisms that enter groundwater as a result of pollution incidents and die because the required conditions are lacking (the 'visitors'), there are two other broad groups; those that are introduced from elsewhere, find the conditions are suitable and thrive, at least for a time (the 'immigrants'); and those that have been present in the environment for a very long time, perhaps since the early alteration of the sediments (the 'residents'). Micro-biologists refer to organisms that are normally resident in the aquifer as indigenous or autochthonous, and to those that have come from elsewhere as allochthonous.

BACTERIAL LIFESTYLES

Bacteria, like people, have several requirements for life. These include water, carbon, a source of energy, and a means of converting that source into energy. For us, the source of energy is food, and the means for its conversion is respiration; we use oxygen from the air to oxidize the food, liberating energy in the process. (The respiration referred to here is a chemical reaction taking place within our cells; it should not be confused with the process of breathing, though of course it is this process that provides oxygen for the reaction.)

Bacteria have a simpler lifestyle. But, like all organisms, their life is a series of chemical reactions that take place in an orderly sequence. These reactions are controlled by **enzymes**, which act upon a **substrate**. An enzyme is a protein that acts as a catalyst, speeding up or enabling a reaction to take place without itself being consumed. The substrate is the material outside the bacterium on which the enzymes act; in essence, bacteria live in or on their food.

In simple terms we think of oxidation as a reaction in which oxygen combines with another substance. More generally, oxidation is used to describe a reaction in which an atom loses one or more electrons: for example, the oxidation of ferrous iron (Fe^{2+}) to ferric iron (Fe^{3+}) by the removal of an electron. To achieve this there must be a substance to accept the electron(s); the electrons may move from one substance to another, but the one that finally retains them is called the **terminal electron acceptor**.

The process of extracting energy from the 'food', viewed at its most basic, involves the transfer of an electron from the food molecule (the **electron donor**) to the electron acceptor. The liberation of the electron by the electron donor involves a release of energy which the organism uses or stores for later use.

The preferred terminal electron acceptor for most organisms is oxygen: this is **aerobic respiration**. Some species of bacteria use other electron acceptors, such as nitrate, sulphate or carbon dioxide. This is **anaerobic respiration**. Some bacteria can respire only aerobically, and are termed **aerobes**. Others can respire *only* anaerobically, and are called **obligate anaerobes**. The use of sulphate as the terminal electron acceptor leads to the process of sulphate reduction; the bacteria which carry it out are termed sulphate-reducing bacteria (SRBs) and are obligate anaerobes. Their presence can lead to the production of hydrogen sulphide (H_2S). If nitrate is used as the electron acceptor, the nitrate is reduced and the end products may be nitrogen dioxide, nitrous oxide, or gaseous nitrogen. The bacteria that reduce nitrate are called denitrifiers, and they can respire aerobically or anaerobically depending on conditions: bacteria that have this useful facility are called **facultative anaerobes**. In general, they will therefore act to remove nitrate only in the absence of oxygen. Bacteria that use carbon dioxide as the terminal electron acceptor are methane producers.

In general, oxygen will be used first, then nitrate, then sulphate, and finally carbon dioxide. As oxygen is consumed, facultative anaerobes may switch to using nitrate, but the organisms that use sulphate and carbon dioxide are specialized and cannot use any other electron acceptor. It is interesting to note that wells in confined aquifers some distance from the outcrop area (e.g. the Chalk in the centre of the London Basin) produce water containing methane, suggesting that the full sequence of activity is present.

All organic compounds can act as electron donors; a good example is glucose, which is easily broken down. As glucose is not readily available in aquifers, bacteria have to find another source of energy. Bacteria that use organic carbon compounds as their food supply will also usually use these compounds as sources of carbon. Those that use inorganic compounds as sources of energy use carbon dioxide as a source of carbon. Inorganic compounds used as electron donors (energy sources) include ammonia, nitrogen dioxide and inorganic sulphur compounds.

Allochthonous bacteria can enter aquifers in recharge water or as a result of human activities. The drilling of a well, for example can provide opportunities for several species to enter. The entry of pathogens is an obvious cause for concern, and if the well is to be used for public supply it will usually be disinfected to reduce the risk of disease. But even non-pathogenic bacteria can cause problems because of their varied 'diets'. Some species attack well linings, others lead to encrustation that can block the openings of screens. The introduction of oxygen into formations where pyrite is present can lead to the generation of acidic waters with high concentrations of sulphate (p. 176), a process which bacteria can accelerate. Sulphate-reducing bacteria can lead to the generation of hydrogen sulphide, which imparts a strong odour of rotten eggs to water, and which is both toxic and explosive at quite low concentrations.

But bacteria can be beneficial as well as harmful. Nitrate pollution of aquifers (Chapter 13) is a major cause of concern in Britain and other countries. One interesting feature is that wells in some confined aquifers yield water that contains tritium in concentrations that imply recent recharge, but which does not contain significant levels of nitrate, although the recharge areas are intensively farmed. Such aquifers include the Lincolnshire Limestone in England and the Chalk of the Paris Basin. It is highly probable that the denitrification that must be taking place to explain this anomaly is caused by bacteria. Such reactions are said to be **biologically**

mediated. Because the pores of the Chalk and the Lincolnshire Limestone are even finer than those of sandstones, most of the bacterial activity is believed to take place on or close to the walls of fissures, probably in thin layers called biofilms.

It is difficult to distinguish absolutely between the effects of micro-organisms and inorganic chemical processes. We can take samples from an aquifer and show that the requisite bacteria are present. We can culture bacteria in the laboratory, creating an artificial environment called a **microcosm**, and show that they achieve the breakdown of, say, nitrate. However, there is still an assumption to make in saying that because the bacteria are present in the aquifer, and because those bacteria reduce nitrate when cultured in the laboratory, they are therefore reducing nitrate while they are in the aquifer; it is possible, for example, that they are present in the aquifer in dormant forms that become active when they are placed in the microcosm. We also have to take great care to be sure that we do not introduce the bacteria as a result of the drilling and sampling. There is great scope here for multi-disciplinary studies.

There is also debate about whether the bacteria in the aquifer have all the requirements necessary to carry out the process; for a long time, for example, it was argued that many aquifers did not contain sufficient carbon for the bacteria to function at the necessary rate. The balance of evidence seems to be that many of these processes could not take place as quickly as they apparently do without the intervention of micro-organisms. At a site in the Chalk near Douai, for example, French scientists have discovered that nitrate concentrations decline at the rate of 0.4 mg/l for every metre that the groundwater flows through that part of the aquifer.

Bacteria can also deal with other pollutants, including organic substances such as oils and industrial solvents, and some pesticides. There is evidence that some breakdown of organic pollutants by bacteria has taken place in some aquifers. However, it appears that it usually takes a long time for the microbes present naturally to adapt to using the pollutant as a source of energy or of carbon.

In attempts to use the potential power of micro-organisms more effectively, groundwater contaminated with some pollutants has been pumped from wells, passed through vessels called bioreactors that contain colonies of appropriate bacteria to break down the pollutants, and re-injected into the ground. Current research is going further, and seeking ways to encourage the formation of bacterial colonies in contaminated environments to clean the water without bringing it to surface. Generally these involve injecting appropriate nutrients into the ground to stimulate the growth of the existing bacteria; recent work, particularly in North America, has shown that it may be well worth while giving them a helping hand, and there seems every prospect that cleaning up contaminated aquifers using the processes of **bioremediation** will become big business.

One other way in which we can put micro-organisms to work for us is to use them as tracers. There is a group of viruses or virus-like agents called **bacteriophages**, or sometimes simply phages, whose natural function is to attack bacteria. Each species of bacteriophage attacks only a single species of bacterium. Bacteriophages can be used as tracers by introducing a culture of a specific bacteriophage into the system to be traced, then sampling the system and introducing the samples to cultures of the host bacterium for that bacteriophage. The presence and concentration of the bacteriophage in the samples can then be expressed in PFU/ml.

Because of the small size of bacteriophages (typically around 0.05 to 0.15 µm for those used in water tracing), they can travel through all but the smallest pores. They can also be introduced into an aquifer in very large numbers – concentrations of 10^{15} PFU in a litre of culture are possible. Theoretically, the identification of only one bacteriophage in a sample would confirm a positive result for the tracer experiment, making bacteriophages very sensitive tracers. They do, however, appear to be absorbed or adsorbed in the rock matrix, so large proportions of the introduced tracers are usually 'lost'. They have been used successfully to trace movement through the Chalk over a distance of 3 km. Because they are themselves a form of microbe, they are useful for simulating the movement of contamination, such as sewage, that contains micro-organisms.

HOW SAFE IS THE WATER?

This question was once asked by people in Britain and North America only when they were travelling to what they regarded as less-developed parts of the world. Recently it has become a major topic of concern to many people at home as well as abroad.

Water from any source must suit the purpose for which it is intended. Water for drinking must be free from pathogenic organisms; in other words, it must have suitable biological quality. Its chemical and physical quality must also be suitable. In chemical terms it must not contain any dissolved or suspended material that would be injurious to health or that would give it an unpleasant taste. Physically it must be at a suitable temperature and not have objectionable colour or cloudiness. To safeguard consumers, various agencies lay down strict requirements that must be met by water-supply undertakings.

In formulating the requirements, the agencies take into account known facts about toxicity of various substances. Some substances have an immediate detrimental effect. Others may be cumulative poisons, collecting in the body over a long period before causing harm; lead is a good example. When deciding on safe limits for these long-term or chronic toxins, the agencies have to allow for the fact that water may not be the only source of the substance in the diet; they may therefore set the permitted level in water at a lower value than would be necessary if water were the sole source to allow for these other components of the daily intake. They also have to consider the potential dangers not only to healthy individuals but also to the most vulnerable of consumers – pregnant women, nursing mothers, babies, small children, and the sick and elderly. The World Health Organization has a set of guidelines that are followed by many countries; the latest edition was published in 1993, replacing a previous version published in 1984. Until 1985 British water undertakings were required to comply with the WHO European Standards for Drinking Water.

In 1980 the European Community published its Directive (80/778/EEC) relating to the quality of water intended for human consumption. All member states of the EC were required to bring into force legislation or administrative provisions to ensure that the Directive was complied with by 1985. In some cases the requirements of the EC Directive were stricter than those previously in existence in member states, and some countries – including Britain – had problems in complying in full with the letter of the new regulations. Examples of some of the requirements are given in Table 1.2. Some of the limits were set for substances that were known or suspected to be harmful; in other words they were based on toxicological grounds. Other limits were more emotive. Two of the limits have caused particular comment: these are the limits set for nitrate and for pesticides.

The concern about nitrate arises from the fact that nitrate in drinking water has been linked to a condition known as **blue baby syndrome** or infantile methaemoglobinaemia. This arises in babies usually less than three months old; the haemoglobin in the blood, which carries oxygen around the body, is converted to methaemoglobin and cannot perform its function. The condition is actually caused by **nitrite** (NO_2), which is formed by the reduction of **nitrate** (NO_3) by bacteria in the intestines. The infant takes on a bluish tinge, particularly around the lips. In severe cases the illness can be fatal, but to put it in context the last reported case in Britain was in 1972, and did not involve a public supply. Indeed, some workers have commented on the absence of the condition in infants consuming water from public supplies with high nitrate levels.

On the other hand there does seem to be a link between the illness and the use of groundwater, a fact which has led to the condition sometimes being referred to as 'well-water cyanosis'. It has been suggested that water from shallow wells contaminated by micro-organisms may have led to diarrhoea, making the babies more vulnerable to the condition; this could explain why the illness seems to be linked with private wells rather than public supplies. The infant affected in 1972 had drunk bottled milk prepared using groundwater from a private well. The water had a high nitrate content, but it is not clear whether the bacterial conversion of nitrate to nitrite arose at least in part in the feeding bottle because of inadequate sterilization.

Table 11.2 Quality of water for human consumption (concentrations are expressed in mg/l)

Parameter	EC Guide level	EC MAC
Calcium	100	–
Magnesium	30	50
Sodium	20	150
Potassium	10	12
Nitrate	25	50
Chloride	25	–
Sulphate	25	250
Aluminium	0.05	0.2
Iron	0.05	0.2
Lead	–	0.05
Arsenic	–	0.05
Mercury	–	0.001
Cyanide	–	0.05
Fluoride	–	1.5
Pesticides (individual)	–	0.0001
(total)	–	0.0005

The pH must be in the range 6.5–8.5
MAC = maximum admissible concentration
The values are based on the EC Directive on the quality of water intended for human consumption (Directive 80/778/EEC). The values are currently (1995) under review

There have also been suggestions that nitrate is linked with cancer, but the evidence for this is more tenuous than for blue baby syndrome. The EC Directive sets a limit for nitrate of 50 mg/l as a maximum admissible concentration (MAC) – the maximum value that is permitted in a public water supply at any time – but suggests a 'guide level' of 25 mg/l. The limits previously set in Britain (based on the WHO standards) were a maximum acceptable limit of 100 mg/l and a maximum desirable limit of 50 mg/l, with a proviso that when concentrations exceeded the desirable limit, local physicians should be advised. These limits compare with that of 44.3 mg/l set in the United States by the Environmental Protection Agency (EPA). (The odd-looking figure of 44.3 arises because nitrate concentrations can be expressed in two ways, in terms of nitrate or of the amount of nitrogen in the nitrate; the EPA figure is based on a concentration of 10 mg/l expressed as nitrogen.)

The EC limit for nitrate in drinking water is therefore not particularly strict in comparison with limits set elsewhere, but it has to be seen in context. As we shall see in Chapter 13, nitrate levels in groundwater have been rising in many parts of Europe and elsewhere, almost certainly as a result of changes in the way farmland is used. To meet the new standards water suppliers have three choices: abandon the worst-affected sources and replace them with new ones with lower nitrate levels (if any can be found); blend the affected water with water that has less nitrate; or treat the water. All of these options are expensive. The last option is difficult because nitrate ions are not easily removed from solution; the most practical technique, that of ion exchange, is expensive to install and operate and produces large quantities of effluent that have to be disposed of. Given the medical uncertainty over the need for such strict limits on nitrate in drinking water, it is not surprising that the issue has generated some heated debate as water companies have been forced to spend many millions of pounds in complying with those limits.

The limits for pesticides set by the EC Directive are much stricter. They are based on the premise that drinking water should not contain pesticides. However, it is impractical to require that a substance should be completely absent from water, because this would require that the instruments used for analysis should be capable of detecting infinitesimal concentrations. Instead, the EC Directive sets a limit that was based on the lowest concentration of pesticide that could be measured by the analytical instruments available at the time. In effect, the Directive said that if the concentration was high enough to be detected it was too high; for this reason the MAC for pesticides is often referred to as a 'surrogate zero'.

In the time since the limits were set, the sensitivity of instruments and techniques has improved to the point that these and lower concentrations can be detected, leading to calls from some quarters that the limits should be lowered again. On the other hand, many people point to the fact that, unlike the WHO and US EPA limits, which are based on toxicological data for individual pesticides, the EC limits treat all pesticides in the same way, regardless of their relative toxicities. The need to comply with the Directive and to allay public concern about pollution from pesticides (Chapter 13), has led some water companies to install advanced treatment using activated carbon, to remove the minute traces of pesticides that occur in groundwater from some sources.

Although it is becoming clear that some species of bacteria are able to live and thrive in some aquifers, groundwater is usually free from pathogenic micro-organisms (pp. 190–1). The filtration treatment given to many surface-water supplies usually happens naturally within aquifers, and groundwater is usually biologically safe. Only if the water moves rapidly through relatively large openings such as fissures or the pores of coarse gravel, or if there is a source of pollution close to the well, or if the well construction permits the entry of contaminated surface water, is abstracted groundwater likely to contain pathogenic organisms. One or more of these factors was probably responsible for the Broad Street epidemic.

Since 1937, when a typhoid epidemic in Croydon was attributed to contaminated water from a Chalk well, public water supplies in Britain have been disinfected. This precaution, which is now a legal requirement, is usually done by adding chlorine at a dose of 1 milligram of chlorine per litre of water; this dose is sufficient to leave a trace of chlorine (the chlorine residual) in the water to protect it from any bacteria that might on rare occasions enter the distribution mains or pipework. Groundwater normally requires little else in the way of treatment.

At least 43 people died in the Croydon epidemic. In the next 50 years, 34 outbreaks of water-borne diseases were reported in the United Kingdom, in which more than 11 000 people were affected but only six died. Of these outbreaks, 21 involved public supplies, of which 11 were contaminated at source, mostly because of the failure of chlorination. None of the six deaths was due to contamination of a public supply at source. A general point to note is that public water supplies are very rarely defective; when they are, by their very nature, they are likely to affect a large number of people.

Because it is less likely to contain pathogenic organisms, groundwater is much more suitable than surface water for supplies of drinking water in isolated areas, where the treatment undertaken in large-scale supply systems is not possible. Provided that proper precautions are taken in drilling and completing wells, groundwater can usually be drunk with no treatment at all. This is a particular advantage in providing supplies in developing countries, where even the most elementary form of water treatment – that of boiling the water for a few minutes to kill the organisms present – is beyond the scope of many of the inhabitants, who cannot afford the necessary firewood or other fuel.

Water for irrigation need not meet such strict biological requirements as drinking water (although it is bad practice to distribute polluted water anywhere where people may come into contact with it), but chemical quality may be important. Industrial requirements vary considerably; some processes need only poor-quality water, while in others the presence of suspended or organic material can block filters, and dissolved material may cause scale formation or may accelerate corrosion. Some processes have their own problems; minute amounts of iron or manganese can be disastrous to a laundry, for example, because they cause staining.

BOTTLED WATERS

Twenty years ago, bottled water was rarely seen in British restaurants and supermarkets. Now it is difficult to persuade some restaurants to serve anything else, and every supermarket has rows of shelves groaning under the weight of bottles and containers of the stuff. Although Britons still consume far less bottled water per head than the French or Germans, British sales have increased tenfold in the past ten years, to more than 500 million litres a year. Many British people who buy it claim that they dislike the taste of tap water or the chlorine residual that it contains, despite the fact that 'blind' tasting panels have usually been unable to distinguish between tap water and non-carbonated bottled waters. But many buy it because they distrust the quality of their tap water. This distrust comes from increased awareness of water-quality issues, arising largely from the publicity given to the nominal failure of some tap water to meet the requirements of the EC Directive, particularly as it relates to nitrate and pesticides. These customers are under the impression that bottled water is in some way purer or safer than the product from their taps, and are prepared to back their belief by paying as much for a litre of water as they would pay for milk or fruit juice – almost as much as they would pay for petrol. Are they right?

The first thing to note is that bottled water is sold in Britain on a totally different basis from water supplied through the taps. It is treated in the same way as foods and other drinks sold in retail outlets, with testing done by local environmental health officers and food safety officers. Bottled water is generally sold in three categories: **natural mineral water**, **spring water** and **table water**. Confusion arises because many bottled waters are sold with a name that includes the word 'spring', giving the impression that the water is derived from a spring and making use of the fact that many people associate spring water with purity. In fact, none of the classes of bottled water usually comes from a spring.

Natural mineral water is groundwater. The only material conditions imposed are that the water must be bottled at 'source' (which is a spring, borehole or well, but in the case of most large bottling plants is almost invariably a borehole) that has been recognized by the local authority for the area in which it lies, that the water must not be treated other than by physical filtration or by the addition or removal of carbon dioxide gas, that it must be free from pollution and that its quality must meet certain standards (which are less strict than those in the Drinking Water Directive). It can be aerated to precipitate iron or manganese before filtration, but it must *not* be disinfected. (It seems faintly ironic that improving the safety of the water renders it unacceptable for this, the highest bottled-water desig-nation.) Because of this last requirement, a hydrogeological assessment is required of the area around the source to demonstrate that it is not vulnerable to pollution, and the water must be monitored for two years before the source can be 'recognized'. The water as abstracted from the ground is usually free from bacteria, pathogenic or otherwise, but tests have shown that many bottled waters contain bacteria at the time of sale. These are generally harmless, and have apparently been introduced during bottling.

'Spring water', as sold in Britain, can be water from almost any source. If it does come from a single groundwater source, that source has not been granted recognition as a 'natural mineral water'; it may not have applied, or for some reason it does not meet the conditions – perhaps because of treatment or perhaps because bottling is not carried out 'at source'. Water sold as 'table water' can be simply tap water that has been carbonated to make it 'fizz', though in some cases it is treated to remove the chlorine residual and other impurities that cause an unpleasant taste. Although generally regarded as much lower than 'natural mineral water' in the hierarchy, 'spring water' and 'table water' do at least have to meet regulations that embody the requirements of the EC Drinking Water Directive.

Many people who live in areas where the tap water is derived from groundwater are nevertheless prepared to pay large sums of money for bottles of groundwater which they carry home from the supermarket. They do this despite the fact that some of the bottled water would not meet the requirements of the Drinking Water Directive, and despite the fact that, because it is not disinfected, it may contain high bacterial counts. Some water companies, no doubt working on the principle that if you cannot beat them you may as well join them, now bottle a tiny proportion of the water from some of their pumping stations; the same water is supplied to consumers in that area at about one-thousandth of the cost of the identical (but untreated) water that goes into the bottles. In the future, new legislation may remove some of the confusion that exists in the minds of many consumers. For the present the business of bottled water remains, in short, a triumph of advertising over education.

GROUNDWATER TEMPERATURE

One of the important physical aspects of groundwater quality is its temperature. Groundwater in the upper few metres of the Earth's crust experiences seasonal fluctuations of temperature. These decrease with depth and generally become negligible, in temperate regions, below 10 m. At that depth the temperature is about equal to the mean annual air temperature (in Britain this is about 10 °C–12 °C) and is remarkably constant.

At greater depths the temperature stays constant with time but increases with depth. This is mainly because heat is being generated by the decay of radioactive minerals – the Earth is acting as a natural nuclear reactor. The amount of heat being generated and therefore the temperature rise with depth varies from place to place. In general it is least in old stable areas like the Canadian Shield, and greatest in areas of recent tectonic and volcanic activity, such as Iceland. Britain has a temperature rise with depth – the **geothermal temperature gradient** – which is near the world average of about 25 °C/km.

In most areas of the world, groundwater at high temperature is therefore found only where water has circulated to great depths. Hot springs have been known in some places for centuries and have formed the basis of resorts; those at Bath in south-west England are an example. There, meteoric groundwater derived from recharge on the Carboniferous Limestone of the Mendip Hills is believed to circulate to depths of more than 4 km. It returns to the surface along fault zones, to emerge as the famous springs, with temperatures as high as 46 °C.

The concern for future energy supplies has led to suggestions that the Earth's natural heat (**geothermal energy**) could be exploited. In some areas this is already done. In Iceland, for example, where the temperature of some of the groundwater is so high that it exists as steam, geothermal energy is used to provide space heating for buildings and for horticulture, and for power generation. Effective use is made of geothermal energy in Italy, New Zealand, Japan, the Philippines and parts of the USA – all areas with a history of recent volcanic or tectonic activity. Now schemes are in progress to try to use geothermal energy in more stable areas with lower geothermal gradients.

A problem is that in these areas it is necessary to drill to great depths to encounter high temperatures; at these depths, permeabilities are often so reduced as to limit well yields. Furthermore, the only techniques available for drilling, testing and producing these wells are those of the oil industry. The high cost of these techniques is justified when producing oil; when applied to hot-water production, however, it means that it costs almost as much to produce a barrel of hot water as a barrel of oil. Since the energy content of the water is much lower, geothermal energy is at present only economical in special circumstances.

Nevertheless, geothermal energy is being used for example to heat buildings in and around Paris. In Britain, trial drilling took place in the 1980s in Southampton, on Humberside and in Northern Ireland. In Southampton a geothermal well provides some of the energy for a district heating scheme that serves buildings in the city centre. Whether such schemes prove to be economic will depend to a large extent on how much other fuel costs increase in future years.

SELECTED REFERENCES

Andrews, J.N. 1992. *Noble gases and radioelements in groundwaters.* In: *Applied groundwater hydrology* (eds R.A. Downing and W.B. Wilkinson). Oxford: Clarendon Press.

Chapelle, F.H. 1993. *Groundwater microbiology and geochemistry.* New York: Wiley.

Domenico, P.A. and F.W. Schwartz 1990. *Physical and chemical hydrogeology.* New York: Wiley.

Downing, R.A., R.H. Parker and D.A. Gray 1992. *Geothermal energy in the United Kingdom.* In: *Applied groundwater hydrology* (eds R.A. Downing and W.B. Wilkinson). Oxford: Clarendon Press.

Ehrlich, H.L. 1990. *Geomicrobiology*, 2nd edn. New York: Marcel Dekker.

Ford, D.C. and P.W. Williams 1989. *Karst geomorphology and hydrology.* London: Chapman & Hall.

Freeze, R.A. and J.A. Cherry 1979. *Groundwater*. Englewood Cliffs, NJ: Prentice-Hall. (See especially Chs 3 and 7.)

Galbraith, N.S., N.J. Barrett and R. Stanwell-Smith 1987. Water and disease after Croydon: a review of water-borne and water-associated disease in the UK 1937–86. *Journal of the Institution of Water and Environmental Management* **1**, 7–21.

Hem, J.D. 1985. *Study and interpretation of the chemical characteristics of natural water*, 3rd edn. US Geological Survey Water-Supply Paper 2254. Washington, DC: US Govt Printing Office.

Jones, K. 1994. *Waterborne diseases*. Inside Science Supplement 73. *New Scientist* **143**, 1933.

Neal, C., D.G. Kinniburgh and P.G. Whitehead 1992. *Shallow groundwater systems*. In: *Applied groundwater hydrology* (eds R.A. Downing and W.B. Wilkinson). Oxford: Clarendon Press.

Price, M., T.C. Atkinson, J.A. Barker, D. Wheeler and R.A. Monkhouse 1992. A tracer study of the danger posed to a chalk aquifer by contaminated highway runoff. *Proceedings of the Institution of Civil Engineers, Water, Maritime and Energy Journal* **96**, 9–18.

Rodda, J.C., R.A. Downing and F.M. Law 1976. *Systematic hydrology*. London: Newnes-Butterworth. (See especially Ch. 7.)

Slade, J.S. 1985. Viruses and drinking water. *Journal of the Institution of Water Engineers and Scientists* **39**, 71–80.

Wilkinson, J.F. 1986. *Introduction to microbiology*. Oxford: Blackwell.

Groundwater: friend or foe? | 12

THE PREFERRED OPTION

We have already seen how water enters and moves through aquifers and other rocks, how we can calculate the amounts of water involved, and how we can construct wells to remove some of that water. But just how do we make use of groundwater? How does groundwater travel from an aquifer to the tap of a consumer? There are many possible ways; a typical British layout is shown in Figure 12.1.

In this arrangement, water is pumped from the well at the **pumping station** to a large tank or reservoir. Not to be confused with a large surface-water storage reservoir, this **distribution reservoir** or **service reservoir** will usually be sited on the highest land in the area which it is to serve. From this high point, water has sufficient head to flow by gravity through the distribution mains to the surrounding towns or villages.

Distribution reservoirs usually appear as rectangular humps surmounting

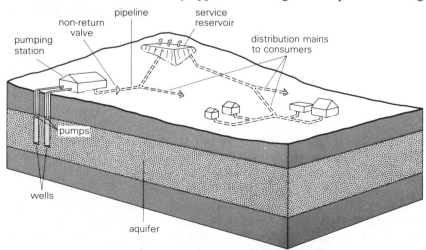

Figure 12.1 A water-distribution system in an area supplied from groundwater sources.

hills, though they are often made as inconspicuous as possible. In flat areas the more conspicuous water towers are a familiar sight; they serve the same purpose of providing a head of water for local distribution, and a constant head for the borehole pump to work against.

Distribution reservoirs provide a small element of storage, each usually holding a volume of water equivalent to about a day's normal supply for the area which it serves. This means that if a pump fails, consumers will still receive water, and there is an adequate reserve to cope with demand at peak times and for emergencies like fire-fighting. This way the well pump can be smaller than it would need to be if it had to cope with all eventualities.

Water engineers are prudent people. Not only do they nearly always have more wells and pumps serving each area than are strictly necessary, they usually arrange that the operating pump can supply all the water necessary by pumping for less than 24 hours each day. Sometimes the pump operates 'on demand', being switched on automatically when the level in the reservoir falls below a pre-set limit, and switched off when the reservoir is full. More often this arrangement is combined with a time switch, so that as much pumping as possible takes place at preferred times (often at night, to use cheap electricity).

In areas underlain by an aquifer, the system outlined above can be very convenient. Wells can be located in or near the areas they serve, avoiding the need for costly pipelines. If demand increases, additional wells can be sunk, so the water-supply system can keep pace with the demand.

This convenience makes groundwater cheap to develop. The capital costs of boreholes and a pumping station are usually much lower than those of a reservoir or a river intake with the same output. Furthermore, because groundwater usually needs little treatment other than routine disinfection, the running costs of a groundwater source are usually much less than those of equivalent surface-water sources, which usually need complex treatment processes to filter and clarify the water.

Cost factors have therefore generally made groundwater the 'preferred option', where it is available with suitable quality, for water for public and industrial supplies in many parts of the world. In most European countries, for example, groundwater provides the majority – in some cases almost all – of the water used for public supplies (Figure 12.2). (In comparing the percentages shown in Figure 12.2 it must be noted that different countries may classify abstractions in different ways: many European countries, for example, classify water abstracted from springs as groundwater, whereas in Britain it is usually classed as surface water.) Similarly, many parts of the United States rely heavily on groundwater (Figure 12.3).

This same convenience can lead to problems. If water is taken from a river catchment then, whether the water is taken from the river or the ground, there will be less water leaving the catchment naturally. This may manifest itself as a reduction in underflow (p. 145), but will usually mean less water flowing in the river. This may seem obvious, but was not always understood.

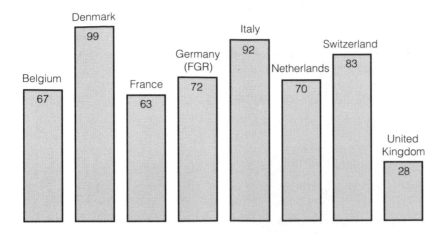

Figure 12.2 Groundwater use as a percentage of public supply: European examples (1986).

If groundwater is used for public supply, or for some industrial processes, it is used but not consumed. Much of the water supplied to our homes is used for washing, bathing and flushing of lavatories. If this water is treated and disposed of to the river whose catchment it was taken from, there need be little net reduction in streamflow. This contrasts with water used for spray irrigation of crops, for example, much of which is lost from the catchment by evapotranspiration.

Problems arise even with non-consumptive use when the water is transferred to another catchment, or treated and returned to the river a long way downstream from its abstraction. Then the flow in the river upstream of the sewage outfall will be reduced. This may not be noticed for some time, because the effect will usually be gradual. It may manifest itself as the less frequent appearance of a winterbourne, or the migration downstream of the perennial head of a river (Chapter 8). For reasons like this there has to be a limit to the amount of water that can be taken from an aquifer in a particular area and many countries now have some form of regulation or licensing system to control abstraction of groundwater or, indeed, water from any source. In England and Wales the current legislation is contained in the Water Resources Act of 1991, and the agency charged with implementing the legislation is the National Rivers Authority (NRA); the NRA is set to become part of a new Environment Agency in 1996.

WATER MANAGEMENT AND CONJUNCTIVE USE

So what happens when the demand for water is in an area not underlain by an aquifer, or when the aquifer is fully exploited and further abstraction

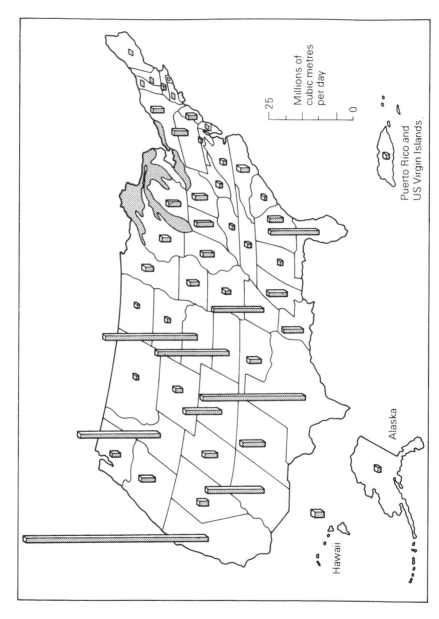

Figure 12.3 Groundwater withdrawals in 1980 for the United States, Puerto Rico and US Virgin Islands. (Source: United States Geological Survey.)

could lead to environmental problems? In these cases, how can water engineers meet the increased demand for water that results from population growth, from increased industrialization, or simply from the greater consumption of water that goes with an improved standard of living?

London provides a good example. The aquifers beneath London (principally the Chalk) are fully developed. Similarly, the natural flow of the Thames is already exploited to the full in dry weather. Large areas of land, especially to the west of London, have been used to create storage reservoirs which can be replenished from the river during high flows; it would be difficult and expensive to build more reservoirs.

One possibility is to use the storage space that has been created in the aquifers below London as a result of the lowering of the potentiometric surface; suitably treated water can be pumped into the aquifer during times of surplus river flow, and pumped out again during periods of scarcity.

This technique, called **artificial recharge**, is being implemented in North London in the valley of the River Lea (a tributary of the Thames) and in the Enfield–Haringey area. The recharge water will be mains water, pumped into the aquifer during winter. Studies of the same technique are also taking place in South London.

Another possibility is to bring water from elsewhere. About 2700 years ago Sennacherib, King of Assyria, built his palace near Nineveh in the north of present-day Iraq. Nineveh was on the River Tigris and so not short of water, but the Tigris carries large amounts of silt and the muddy water was not good enough for the conqueror of Babylon. Sennacherib therefore had a canal constructed to carry clear water some 55 km from the mountains north of Nineveh; the canal included a stone aqueduct 300 m long across a valley. He set a trend that has been followed by many subsequent civilizations from the Romans to those of the present day. The water supply of Birmingham, for example, is carried nearly 120 km by an aqueduct from a group of reservoirs in Mid-Wales, formed by damming two tributaries of the River Wye.

But taking water by aqueduct or pipeline is costly, because the structures are expensive to build. It also means that the water serves no useful function between the source and the place being supplied (the **demand**). For these reasons, water engineers devised the technique called **river augmentation**. Instead of using an artificial channel, the water is put into a river and allowed to flow naturally to the area of demand, where it is abstracted by a conventional river intake works.

In this way the water is added to the flow of the river for navigation, recreation, fisheries and sewage dilution – unlike water conveyed by pipeline, which cannot serve any of these functions. And these benefits can be made available at the very time they are most needed, when the natural flow of the river is low, for it is usually then that augmentation is at its greatest. A disadvantage is that the water, having been in the river, may need more treatment than it would if it had been piped direct from a reservoir.

River augmentation can also be practised using groundwater as the source. In Britain, 13 schemes involving the abstraction of groundwater to support river abstraction further downstream are either complete or in various stages of development. In addition, 11 generally smaller schemes are completed or planned to increase the flows of streams that have been depleted by groundwater abstraction. All of these schemes are intended to operate only in periods of dry weather, when natural flow in the rivers is unacceptably low, either for abstraction or to maintain the health of the river.

To work successfully these schemes require careful planning, particularly in the siting of the production wells. If these wells are too close to the river or its tributaries they simply intercept groundwater that would have flowed to the river anyway. A measure of this effect is the **net gain**, which is defined as the increase in flow of the river divided by the amount of water pumped into the river; it is usually expressed as a percentage. Initially the net gain will be 100%, but it declines as the cones of depression from the abstraction wells spread out and intercept flow to the river.

In formations like the Chalk and Jurassic limestones, the permeability is so great as a result of fissuring, and the specific yield is so low, that the cone of depression spreads quickly through the aquifer and the net gain soon declines. This caused problems with Britain's first river augmentation from groundwater, which used water from the Chalk in the Berkshire Downs to increase the flow in a tributary of the Thames; the first wells were too close to the river. In sandstones, on the other hand, the permeability is generally lower and the specific yield higher. The cone of depression therefore grows more slowly, and more water can be removed before the flow of the river is seriously affected.

These schemes are examples of the **conjunctive use** of groundwater and surface water. One feature common to most conjuctive-use schemes is that they use groundwater storage to augment surface-water supplies in dry weather. This usually results in the water table being lowered below its normal minimum level, but this is not a disadvantage: the extra volume dewatered represents storage that would otherwise not be used (Figure 12.4). This storage is usually quickly replenished in winter, the only difference being that it takes longer for the water table to return to its normal maximum and therefore longer for all the springs and streams to begin flowing again. This is an advantage: as there is generally a surplus of water in winter, this flow is really a waste of water and may even be a nuisance. If the scheme is being managed properly, this reduced winter flow need not be seen, as it sometimes is, as the first stage in the 'drying up' of the aquifer.

GROUNDWATER AS A PROBLEM

It is convenient when we require a water supply to be able to drill a well into an aquifer, insert a pump, and abstract water. What is less convenient is that groundwater will flow into any excavation in the saturated zone, unless we somehow prevent it. Whether it is a mine shaft, a gravel pit, a railway tunnel, or simply the basement of a house, if it is within the saturated zone it is going to be affected by groundwater.

The effects do not consist merely of the entry of water; the presence of groundwater can also affect the strength and stability of the ground. Anyone who has tried to dig a hole in a sandy beach will have experienced both of these effects. When a hole is dug into firm damp sand on a beach, the sides of the hole, to begin with, stand without being supported. As soon as the hole reaches the level of the sea or lake, however, we strike water – we have in effect reached the local water table. At that point, usually, water and sand flow in together, and the bottom of our hole turns into a wet sandy mess. There in miniature are the two problems that groundwater can cause in excavations – flooding and instability.

One person who discovered the problems that groundwater can cause for anyone working below the water table was Guy Fawkes, the mercenary hired by a group of disgruntled Catholic noblemen to dispose of King James I in 1605. The group plotted to do this by the simple if bloody expedient of blowing up the House of Lords when the King opened Parliament on 5 November that year. Their first idea was to dig a tunnel by hand

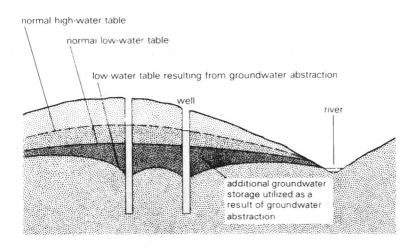

Figure 12.4 Extra groundwater storage utilized in a conjunctive-use scheme.

from the house of one of the ringleaders, Thomas Percy, towards the cellar of the House of Lords, and pack it with explosives.

The tunnel was in alluvial deposits very close to the River Thames, and not surprisingly groundwater began to seep into part of it. The conspirators were forced to abandon this plan, and opted for the simpler alternative of renting part of the cellar, ostensibly to store fuel for the winter. It was here that Fawkes was discovered, early in the morning of 5 November, with a tinder box and slowmatch, and with more than 30 barrels of gunpowder concealed among the logs and coal in the 'fuel store'.

Some historians believe that the Gunpowder Plot was doomed from the outset, with Robert Cecil, Earl of Salisbury and the equivalent of the modern Home Secretary, aware of the plotters' every move. Others believe that it was the sending of a letter by one of the plotters, warning a Catholic nobleman to stay away from the opening of Parliament, that aroused suspicion and led to the search of the cellars. But it is interesting to speculate on what would have happened had that seepage not forced the conspirators to abandon their tunnel and opt for the plan of using the cellar; would British children now be celebrating the heroism of Guy Fawkes every November, instead of burning his effigy? Did groundwater change the course of history?

The extent to which the inflow of groundwater causes a problem is dependent on its quantity, and hence on the permeability of the surrounding rock or soil. The extent to which the groundwater causes instability however is dependent on the pore-water pressure, and need bear no relationship to quantity or permeability; indeed, it is often more of a problem in poorly permeable material than in highly permeable rocks.

GROUNDWATER AS A CAUSE OF INSTABILITY

We can return to the inflow problem in a while. First let us consider the question of instability.

We saw in Figure 7.3 that the weight of a confining bed is supported partly by the aquifer framework and partly by the water in the pore space of the aquifer. Actually, these considerations apply to any level of any porous rock unit. Suppose we consider the forces acting on a horizontal plane at some arbitrary depth (Figure 12.5). The pressure acting downwards on this plane is the weight of the overlying material divided by the area of the plane. This is balanced by the upward pressures on the plane, which are the grain-to-grain contact pressure w, and the pore-water pressure p (above atmospheric), i.e.

$$W = w + p. \tag{12.1}$$

The study of these and similar forces within consolidated rock is termed 'rock mechanics'; within unconsolidated material it is called **soil mechanics**.

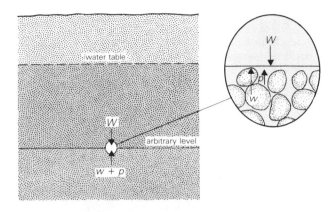

Figure 12.5 Total stress and effective stress. At any level in the ground, the pressure W (the total stress) exerted by the overlying material is balanced by the average grain contact pressure w (the effective stress), and the pore-water pressure, p. (p is measured relative to atmospheric pressure.)

In soil-mechanics terms, the total pressure W is called the **total stress**; the portion w that is supported by grain-to-grain contact is termed the **effective stress**. In many situations in hydrogeology the total stress W does not change, because the weight of the overlying material is unaltered. What *can* change however are the relative sizes of w and p, i.e. the proportions of the

RUNNING SAND AND QUICKSAND

When groundwater flows horizontally through an aquifer, there is no change in elevation head. Assuming that the velocity head is negligible, all of the head loss associated with the flow is therefore caused by the reduction in pressure head. This means that as the water flows through a small block or element of the aquifer, the pressure exerted by the water on the upstream side is greater than that exerted by the water on the downstream side (Figure 12.6). This in turn means that the water is exerting on each element of the aquifer a net force, which is trying to push it along in the direction of the flow of groundwater. This force is called the **seepage force**.

If the aquifer is an unconsolidated gravel or sand or silt, the pressure will be greater on the upstream side of each grain than on the downstream side, and it is relatively easy to imagine the seepage force trying to push the grains along. In the absence of any cement or cohesion, the movement is resisted by the shear strength of the sand arising from the effective stress, and by the grains downstream. There will therefore usually be no mass movement of aquifer material, though very fine particles may be dislodged and carried through the pore spaces of predominantly coarser-grained materials.

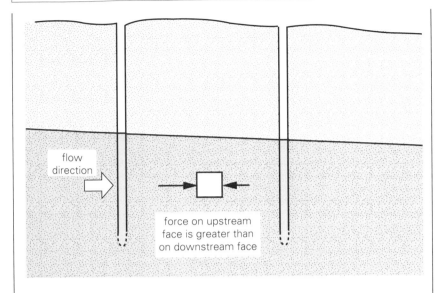

(a)

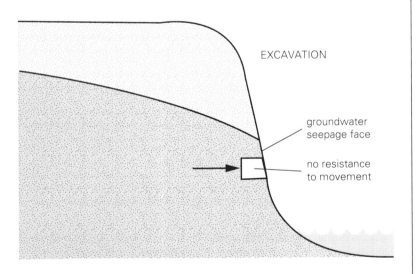

(b)

Figure 12.6 Seepage force caused by groundwater flow. (a) When groundwater flows through an aquifer, the energy loss means that the water exerts a slightly greater force on the upstream face of each element of aquifer than on the downstream face. Normally this force is resisted by the grains downstream and by the shear strength of the material. (b) In the side of an excavation there are no grains downstream to resist movement, and the seepage force may be great enough to overcome the shear strength. This will be especially true if the shear strength is reduced by a high pore-water pressure.

If the material possesses no cohesion, and the effective stress is reduced to zero, then the movement of the grains is resisted only by the grains downstream. If that support is removed, as it will be, for example, if a trench is dug through the ground, then there will be nothing to prevent movement. The material will begin to move into the trench (Figure 12.6b).

In practice, it is unusual for the effective stress to be reduced to zero in such conditions, because the flow of water into the trench will lower the pore-water pressure. What may happen is that the effective stress is reduced to the point where the shear strength of the formation is overcome by the seepage force. This is most likely to happen where the hydraulic gradient is steep, because then the pressure difference across each grain will be relatively great. Steep hydraulic gradients are most likely to arise in material of low permeability, such as poorly-sorted fine sands or mixtures of sand and silt, which will have little cohesion. These materials are therefore most likely to be the ones that 'run' into an excavation. If a layer of such material is overlain and underlain by coarser, more permeable sand, it will be the fine material that moves because there will be a steeper hydraulic gradient through it and therefore greater seepage forces. Drillers and earthworks contractors often refer to such a layer as 'running sand', as though it were a particular type of material, when in reality it is the hydrogeological setting as much as the rock type that is responsible for the movement.

Movement of fine material in response to seepage forces is referred to as **piping**, and it can potentially be a major problem for civil engineers constructing embankments and earth dams.

If there is a vertical component to the flow of groundwater, then there will also be a vertical seepage force. If the flow is downward, then the force will be downward, and in an unconsolidated sand or silt this seepage force will augment the weight of each grain and tend to press the grains together. This is one reason why garden soil compacts if you apply too much water too quickly when watering. If the groundwater is flowing upwards, there will be an upward seepage force that will help to offset the weight of each grain. Where ground-water is discharging naturally and in a localized manner through fine materials – as in the bed of a stream or on a beach – the upward seepage force may be strong enough to overcome the weight of the grains, so that the silt behaves more or less as a liquid, with little or no strength to support anything placed upon it. This is the **quicksand** condition, much quoted by writers of adventure stories.

total stress carried by the framework and the water. As we saw in Chapter 7, if p is reduced then w must increase, and the aquifer may be compacted as a result of the increased effective stress, leading to subsidence.

If the total stress W is reduced for some reason – perhaps by excavation of some of the overlying rock or soil – or if p is increased, then the effective stress w will be reduced. This means that the grains will not be pressed so hard against each other. If W were reduced or p increased to the point where $W = p$, then all the weight of the overlying material would be borne by the pore water; the effective stress would be reduced to zero, because there would no longer be any significant grain-to-grain pressure.

In consolidated rock such as sandstone the grains are held together by cement. In most unconsolidated materials it is the friction between grains resulting from the effective stress that gives the material most of its **shear strength** – its resistance to movement of the particles across each other. In the absence of any effective stress, there is a tendency for the particles of most sediments to stick together slightly, particularly if clay minerals are present. This gives rise to some shear strength, called **cohesion**, even when the effective stress is zero.

Clean sand has little or no cohesion. If the pore pressure were to rise or the total stress fall so that the effective stress was reduced to zero, the sand would lose all its shear strength.

Even if the effective stress is not reduced to zero, the reduction in shear strength caused by a decrease in W or an increase in pore pressure can sometimes lead to material slumping into excavations (as into our hole on the beach) (see Box) or to the occurrence of landslips. Many landslips occur during or at the end of the rainy season, or after exceptional rainfall.

An interesting question is: what happens when the pore-water pressure is less than atmospheric, i.e. when p in Equation 12.1 is negative? In other words, what happens in the unsaturated zone? Use of Equation 12.1 suggests that if p is negative (as it is above the water table) then w is greater than W, i.e. the effective stress is greater than the total stress and the water is helping to hold the grains in contact. It might seem reasonable to suppose that the surface-tension forces in the unsaturated zone do help to hold the grains in contact, just as the water films around sand grains mean that wet sand will cling to a knife blade whereas dry sand will not. The sand in our hole on the beach was stable above the water table and only became unstable where it was fully saturated. However, the stress relations in the unsaturated zone are more complicated than these simple examples might suggest, and in general Equation 12.1 should not be relied upon in the unsaturated zone.

In consolidated rock, movement along joints and faults can be influenced by changes in pore pressure. A particular example is the occurrence of earthquakes. Stresses associated with the movement of crustal plates build up in the rocks at various places in the Earth's crust. The stresses increase until they exceed the strength of the rock, which abruptly fractures. The resulting fracture is called a **fault**; the sudden release of energy causes the **earthquake**. Once a fault has developed – usually at a weak place in the rock – subsequent movements tend to occur along it; stresses build up until they overcome the resistance to movement caused by the friction of the rocks along the fault surface. When the frictional forces – which depend on the effective stress and hence on the pore pressure – are exceeded, an earthquake occurs. This suggests the possibility that we may be able to predict earthquakes by monitoring the fluid pressures along the major fault zones where earthquakes occur. Even better, by injecting water into these zones we may someday be able to control earthquakes. By keeping frictional resistance low, we might be able to prevent large stresses from building up; in this way we could ensure that the stresses were relieved a little at a time, without the dramatic releases of energy that accompany major movements.

In general, studies of pore-water pressure problems and the measures used to combat them are an essential part of the work of the engineering geologist, and reference should be made to textbooks on that subject for a full treatment of them. Suffice to say here that there are few problems of soil and rock mechanics that are not in some way influenced by groundwater, and it is unfortunate that engineering geologists and hydrogeologists often use different terminology to describe the same thing.

One particular example of pore-water pressure as a problem occurs in connection with concrete dams. By its very purpose, a dam causes a considerable difference in head between its upstream and downstream faces. Concrete dams should always be sited on rocks of low permeability but, as we saw in Chapter 6, no rock is completely impermeable and under the influence of the head difference water will flow under the dam. This flow is undesirable not only because of the loss of water, but because it leads to increased pore pressures in the rock beneath. The resulting reduction in effective stress reduces the frictional forces that are preventing the dam from being moved downstream by the pressure of the impounded water (Figure 12.7a).

The seepage of water beneath the dam tends to decrease with depth. To lessen the effects, engineers use two techniques. First, they inject grout (p. 219) into the rock from a line of boreholes drilled beneath the dam. The grout strengthens the rock and reduces its permeability and porosity, leading to a reduction in seepage. Second, they drill another line of boreholes parallel to the grout curtain. These boreholes relieve any build-up of pore pressure by allowing the water to drain into a gallery within the dam, from where it can be removed (Figure 12.7b).

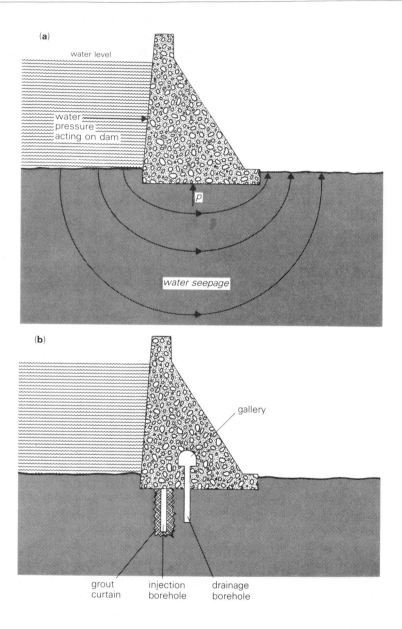

Figure 12.7 Seepage under dams. (a) The head of water behind a concrete gravity dam causes increased pore-water pressure beneath the dam; this in turn decreases the frictional resistance that is preventing the dam from being pushed downstream by the water behind it. (b) A grout curtain reduces the seepage of water beneath the dam, and drainage boreholes relieve the pore-water pressure. A diaphragm wall may be used instead of a grout curtain.

CONTROLLING GROUNDWATER IN ENGINEERING WORKS

The inflow of groundwater into excavations can be prevented or reduced in a variety of ways. The technique chosen will depend on factors like the permeability and the hydraulic gradients present, whether it is also necessary to stabilize the ground, the size of the excavation, and whether control is to be temporary or permanent. Sometimes one technique is used as a temporary control while more permanent controls are being installed; frequently two or more techniques are used in combination. The more common techniques are listed below.

Sheet piling

This is a groundwater-exclusion technique, used where saturated, permeable unconsolidated materials overlie relatively impermeable materials at shallow depth. Corrugated steel sheets are driven down to the impermeable layer from the surface. The method provides reasonable ground support at the sides of excavations, but rarely provides a complete barrier to groundwater movement.

Diaphragm walling

This functions in a similar way to sheet piling, but sophisticated emplacement techniques are used to install a thin concrete wall (the diaphragm wall) through the permeable material. It is more expensive than sheet piling but provides a much more effective barrier to groundwater movement.

Grouting

In simple terms grouting involves injecting into the pore spaces of the rock a liquid (called **grout**) that will harden and set. The method is used for permanent reduction of permeability in all rock types and – especially in unconsolidated rocks – to help stabilize the ground. Grout is injected through specially drilled boreholes, and packers (p. 163) may be used to isolate certain intervals for grout treatment. The two main types of grout are **cement grouts** (which are essentially thin concrete, sometimes containing clay) and **chemical grouts**. The cement grouts are sometimes called **particulate grouts** because they consist essentially of particles suspended in water; because of this there is a lower limit of about 0.2 mm to the size of pore or fissure into which they can be injected, because of the natural filtering action of the rock. Chemical grouts are similar to epoxy-resin adhesives; they do not suffer to the same extent from the filtering problem, and depending on their viscosity may penetrate for considerable distances into the formation. Large volumes of expensive thin chemical grout can be 'lost' along fissures or into permeable layers unless care is taken. The usual procedure is to grout with a cement grout first to seal major fissures or very permeable layers; when this has hardened, successive applications with chemical grouts of reducing viscosity are used to seal the less permeable intervening material. It is impossible, however, to fill completely all the pore space in a rock.

It is important that grouting should be carried out before excavation, because the presence of the excavation will usually cause a hydraulic gradient towards it which may make it difficult to place the grout where it is needed. Detailed investigations, particularly of permeability distribution, are necessary before any grouting operation if costs are to be predicted accurately.

Freezing

The technique of freezing to provide a temporary reduction in permeability dates back to the end of the 19th century. It is most frequently used in association with the sinking of mine shafts. Boreholes, cased throughout their depth, are drilled around the shaft site, and refrigerated brine is pumped into each borehole through an access pipe. The brine is kept in circulation, cooling the rock and freezing the groundwater. A cylinder of frozen ground thus forms around each borehole and these solid cylinders eventually combine to surround the shaft site with an ice barrier. The excavation of the shaft can then proceed within the protection of the frozen ground. Once the shaft is completed and lined, the ground is allowed to thaw.

Freezing is expensive, but the costs can be predicted with reasonable accuracy; grouting costs on the other hand can escalate if the permeability variations are not well known in advance. Freezing will usually provide the only practical means of dealing with large thicknesses of saturated fine-grained materials, which are difficult or impossible to grout.

Compressed air

In tunnelling operations, the entry of groundwater into the unlined portion of the tunnel is sometimes reduced by artificially increasing the air pressure – typically to about twice normal atmospheric pressure – until the permanent waterproof lining can be installed. The increased pressure reduces the head difference between the rock and the tunnel, and therefore reduces the inflow.

This technique means that the workmen must work in an artificially high air pressure, entering and leaving through an air lock. If the pressure is significantly above atmospheric strict precautions have to be taken, including provision of depressurization facilities similar to those used by deep-sea divers.

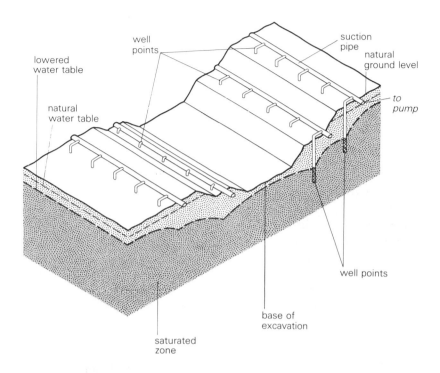

Figure 12.8 Site dewatering. A two-stage dewatering scheme, using well points to lower the water table below the base of an excavation.

Groundwater lowering

This technique, also known as **dewatering**, involves abstracting groundwater so as to lower the water table and permit excavation or other work to proceed within the dewatered area (Figure 12.8). The dewatering may be temporary – perhaps for the construction of a shallow road or railway tunnel which will eventually have a waterproof lining – or permanent. Permanent lowering of the water table is used so that buildings can be constructed in areas such as river flood plains, where the water table is close to ground level.

In confined formations groundwater may be abstracted to reduce the head and hence the inflows into underground openings such as shafts. The technique can be particularly effective if used in association with grouting.

In theory, dewatering can be achieved in several ways. A common technique for temporary dewatering is to install rows of small-diameter boreholes called **well points** around the site. These are usually driven or washed into place by pumping water into them as they are forced into the ground – a process called **jetting**. When they are in place, several are connected to a common suction pipe and pumped using a surface pump.

Because of the lift limitations of this type of pump, successive lines of well points may have to be installed at different levels within the excavation if the water table is to be lowered more than a few metres.

Drainage

Whatever method is used to reduce the permeability of the rock around an excavation, there is still usually some flow of water into the excavation which has to be pumped out. The usual way of doing this is to drain the water to sumps at the lowest part of the excavation, and to pump it out from there.

If the excavation is in rock of low permeability, then drainage of this kind may be the simplest and cheapest way of dealing with groundwater, without resort to other techniques. It may even be an adequate long-term solution, for example where road or railway cuttings pass just below the water table for part of their length.

Sometimes drainage operations can unintentionally become dewatering schemes, with surprising results. London Regional Transport's District Line passes through cuttings or 'cut-and-cover' tunnels for much of its length. The drainage of the railway has exerted a considerable influence on the groundwater-flow patterns in the gravels north of the Thames, the full effect only being discovered during the course of investigations for the Thames Barrier. Between West Kensington and Temple stations, 6800 m^3 of water is pumped out each day to keep the track dry.

MINE DRAINAGE

Deep mines inevitably have most of their workings below the water table. It would not be practical or economical to line all of the workings to exclude groundwater, but fortunately most mines are in rocks of low permeability and the workings are kept dry by pumping. Metalliferous ores are usually mined from metamorphic rocks. At depth these rocks rarely have open joints or fractures, so that seepage into the workings is slow except in the vicinity of major faults or zones of shattered rock, which can generally be grouted or lined individually.

In Britain, the Carboniferous rocks in which coal occurs are of relatively low permeability, so that it is usually possible to pump water from the workings faster than it can enter. The Coal Measures are effectively separated into hydraulically distinct units by beds of clay. Mine drainage may locally dewater some of these units – depressing the potentiometric surface for that unit by hundreds of metres – while strata above and below remain saturated.

In some coalfields the Carboniferous rocks are overlain by saturated permeable Permo-Triassic sandstones. These sandstones are separated from the workings by less permeable Carboniferous strata, but the access shafts to the mines must penetrate the permeable material. Several such shafts were sunk for the new mine near Selby in Yorkshire, where freezing, grouting and pumping to reduce the groundwater head were used to permit excavation. On completion, shafts like these are lined to exclude water, since pumping alone could not keep them dry. Early shafts were lined with brick or with cast-iron linings called 'tubbing', but modern shafts are generally lined with concrete.

The water pumped from mine workings frequently has a high content of dissolved solids (Chapter 11), including iron compounds. Its disposal may therefore require some care: if merely pumped into surface drainage or onto permeable ground it could contaminate river water or local aquifers, and cause unsightly iron-oxide deposits.

When workings are abandoned, it may be necessary to continue pumping to protect adjacent workings. If pumping from a disused mine is stopped, the mine will become filled with water, and may endanger later underground operations in the area. A terrible example of this occurred in March 1973 at Lofthouse Colliery, Yorkshire, when a new coalface was excavated too close to old flooded workings, abandoned in the 19th century and unknown to the modern mine's surveyors. In the sudden in-rush of water, seven men were killed. The report of the Public Inquiry is a grim warning to make the fullest study of historical records before commencing new works.

RISING WATER LEVELS

A condition that has been causing concern in several parts of the world over the last few years is that of rising groundwater levels. Many major cities in Britain and overseas underwent rapid growth of population and industry in the 19th and early 20th centuries, with a corresponding increase in the demand for water. Where groundwater was readily available beneath such cities it was keenly exploited, and was often a factor in the development of the city. Such exploitation, usually uncontrolled in those days by legislation, led to water tables being drastically lowered, often by many tens of metres.

In London the maximum decline was reached in 1965; in a well beside the National Gallery in Trafalgar Square the water level fell to more than 80 m below sea level. The potentiometric surface was lowered below the top of the Chalk, which changed from a confined to an unconfined aquifer. Large declines also occurred in Birmingham, Liverpool and Nottingham (all on Permo-Triassic sandstones) and in Paris, Tokyo and New York.

A combination of legislation to prevent excessive abstraction, of deterioration in the quality of groundwater in many cities leading to sources being abandoned, and of reduction in demand for water as industries have moved out of city centres, has led to abstraction of groundwater being greatly reduced in many cities. This in turn has led to groundwater levels beginning to recover, often with remarkable speed – as much as 2.5 m per year in London.

The decades during which water levels were depressed were frequently periods of major development, with road and rail tunnels, deep basements and tall buildings being put in place. Many of these developments involved construction in ground which would have been in the saturated zone under natural conditions, but which at the time was often tens of metres above the depressed water table. Not surprisingly, those responsible for the construction often assumed that the water table would remain where it was at the time of their site investigation. They found they were wrong when basements, tunnels and underground car parks either became prone to flooding or required additional pumping to keep them dry.

Other unwanted side-effects have been the remobilization of pollutants, trapped in the unsaturated zone, which have been released as water tables have risen. In New York, oil spilled in the 1950s had apparently migrated down to the depressed water table and lain there undetected. As the water table rose as a result of reduced groundwater abstraction, the oil rose with it until in 1978 it began to be discharged into a tributary of East River.

Other cases of rising water tables are caused not by a reduction in abstraction but by an increase in recharge. The urbanization of an area usually means that there is less recharge to an underlying aquifer, because precipitation is intercepted by buildings and paved surfaces, and is frequently led into sewers for disposal to surface watercourses or else evaporates. This effect contributes to the decline of groundwater levels beneath cities, but is usually less important than the effects of abstraction. In arid areas, however, water is often brought into the city from elsewhere and is applied to the ground to water gardens and parks, the application often being much more than is necessary. Surplus water infiltrates and causes the water table to rise. This has occurred in several parts of the Middle East, especially around the Persian Gulf. The effect is aggravated if the city has an adequate water supply but is not fully sewered, so that household wastewater is returned to the ground via septic tanks.

Leaking water mains also pose a problem. In the late 1980s it was conservatively estimated that about 25% of all water put into the mains in Britain as public water supplies was lost from the system. This amounted to about 80 litres per day for every man, woman and child in the country, or a staggering total of 1400 million cubic metres each year – the equivalent of seven reservoirs the size of Kielder Water. Since then there has been a major effort by water undertakings, encouraged by the National Rivers Authority, to reduce these losses. Beneath a large city, losses on this scale could make the difference between a rising and a falling water table.

The closure of mines and the cessation of pumping can also lead to the rise of water levels. Here the problem is usually that the workings flood, and minerals that have been oxidized by exposure to air during the lifetime of the mine release unwanted oxidation products into the water (Chapter 11). The eventual uncontrolled discharge of this water at the surface can lead to major problems in rivers.

SELECTED REFERENCES

Freeze, R.A. and J.A. Cherry 1979. *Groundwater.* Englewood Cliffs, NJ: Prentice-Hall. (See especially Ch. 10.)

Gray, D.A. and S.S.D. Foster 1972. Urban influences upon groundwater conditions in Thames Flood Plain deposits of Central London. *Philosophical Transactions of the Royal Society of London*, Series A **272,** 245–57.

Hardcastle, B.J. 1978. From concept to commissioning. In: *Proceedings of conference, Thames Groundwater Scheme*, 5–31. London: Institution of Civil Engineers.

Middelboe, S. 1987. Guy certainly was not joking. *New Civil Engineer*, 5 November 1987, 32–34.

Mines Inspectorate 1973. *Inrush at Lofthouse Colliery, Yorkshire.* Cmnd 5419. London: HMSO.

Owen, M., H.G. Headworth and M. Morgan-Jones 1991. *Groundwater in basin management.* In: *Applied groundwater hydrology* (eds R.A. Downing and W.B. Wilkinson). Oxford: Clarendon Press.

Price, M. and D.W. Reed 1989. The influence of mains leakage and urban drainage on groundwater levels beneath conurbations in the UK. *Proceedings of the Institution of Civil Engineers*, Part 1, **86,** 31–39.

Simpson, B., T. Blower, R.N. Craig and W.B. Wilkinson 1989. *The engineering implications of rising groundwater levels in the deep aquifer beneath London.* CIRIA Special Publication 89. London: Construction Industry Research and Information Association.

Somerville, S.H. 1986. *Control of groundwater for temporary works.* CIRIA Report 113. London: Construction Industry Research and Information Association.

Todd, D.K. 1980. *Groundwater hydrology*, 2nd edn. New York: Wiley.

Wilkinson, W.B. and F.C. Brassington 1992. *Rising groundwater levels – an international problem.* In: *Applied groundwater hydrology* (eds R.A. Downing and W.B. Wilkinson). Oxford: Clarendon Press.

<table>
<tr><td>13</td><td># What goes down must come up</td></tr>
</table>

THE MYTH OF PURITY

Ask people what characteristic they most desire in their drinking water, and I suspect that many will reply 'purity'. The concept of pure water is one that is pushed hard by many consumer groups and suppliers of bottled waters, but – as we saw in Chapter 11 – it is a myth. No natural water is really pure, if by 'pure' we mean that nothing is dissolved in it. Rain water contains dissolved carbon dioxide and a little dissolved material – chiefly sodium chloride – carried into the atmosphere from the oceans. It also picks up dust borne aloft from volcanic eruptions, forest fires and wind erosion, as well as the atmospheric pollutants introduced by human activity. Snow is even less pure, because the greater area and slower fall of a snowflake make it more efficient than a raindrop of the same mass at scavenging these substances from the air.

Once the rain or snow has reached the ground and infiltrated, it has even more opportunity to start dissolving all manner of substances from the minerals in the soil and rocks through which it passes. So, although many people regard pure water as something that is desirable because it is 'natural', ironically the only really pure water is man-made. You can tell the people you ask that, if they really want their drinking water to be pure, they should drink distilled water – and very flat and unpleasant they will find it.

KEEP IT CLEAN

But your friends may give a different definition of purity. They may say that 'pure' means that the water is not contaminated or not polluted. Contamination and pollution can be emotive words. Both imply that the water has somehow been rendered unclean or less fit for use. They have different origins, but many people and many dictionaries define them as meaning much the same. Other people try to give them separate meanings, saying for example that both pollution and contamination of water mean that the water contains something it should not, but that one implies that the 'something' is harmful.

We could debate the definitions of pollution and contamination for several pages, and I do not think we should be any further forward at the end. It seems to me that in general usage they are interchangeable, except that we perhaps talk of **contamination** as something that is local or arising from a specific incident, and **pollution** as something that is more widespread. So we say that an individual well or the land around a particular factory or site is *contaminated*, or that an incident released a plume of *contamination* into the atmosphere, but tend to talk in general terms of *water pollution* or *air pollution* when these have arisen from a variety of causes or have moved away from the original source.

Many people regard contamination and pollution as the introduction of something unnatural, but what is natural on this increasingly crowded, 'managed' planet of ours? Cow dung is natural, but few of us would like it in our drinking water. And is the dung still natural if the cattle producing it have been bred by artificial insemination, fed almost totally on manufactured feed, and hardly been outside the intensive livestock unit in which they have been reared?

It is worth noting that things that are natural can be harmful, that not all things that are unnatural will do you harm, and that much depends on the concentration of the substance in question and how much of it you are likely to ingest. Arsenic, for example, is for many people a classic poison and is harmful even at low concentrations, but it can occur quite naturally in water. Nitrate can be introduced artificially, but needs to be present in much greater concentrations than arsenic to do any harm – if indeed it does any harm at all. The prospect of water being contaminated by bacteria also frightens most people, but there are many species of bacteria that pose no threat to human health, and there are many micro-organisms other than bacteria that do.

I shall use contamination to mean that something has been introduced to the environment as a result of human activity, and pollution as a term for the more widespread results of contamination. Thus in the example above, animal dung from a farm represents contamination: dung from a wild animal does not.

The usage that I have suggested does not differ markedly from that given in the European Directive on the protection of groundwater against pollution caused by certain dangerous substances (Directive 80/68/EEC), usually referred to as the 'Groundwater Directive'. In this Directive pollution means the discharge by humans, directly or indirectly, of substances or energy into groundwater, the results of which are such as to endanger human health or water supplies, harm living resources and the aquatic ecosystem or interfere with other legitimate uses of water.

It is also worth noting that man-made pollution is not new. Studies of cores taken from the Greenland ice-cap have shown that the levels of lead in the atmosphere between 500 BC and AD 300 were up to four times the 'natural' level before and after that period. The increase was probably caused by the smelting of galena, a lead ore, by the Greeks and Romans.

Surface waters, such as rivers and lakes, are readily vulnerable to pollution. It is probably fair to say that they have been polluted ever since there have been people on Earth to pollute them. We are so obviously aware of this that we have taken action. So, in the 19th century, this awareness led to the abandoning of rivers as sources of supply to many conurbations; instead, reservoirs were built in upland areas where they could be protected, often by keeping the public completely away from them. Water engineers developed techniques of treating even grossly polluted water, so that almost any source can now be turned into a safe water supply, provided that we are prepared to spend enough money. At the same time we began to protect our rivers and lakes from discharges of untreated sewage and industrial effluents.

Groundwater may not be pure, but it is much less vulnerable to the simple pollution by human and animal wastes that has affected surface waters in and around human settlements. This led to the mistaken assumption that groundwater was largely invulnerable to pollution. In the early days of hydrogeology the emphasis was on quantity and reliability of yield. Provided that the water was drinkable, quality was not considered a major factor.

Since World War II, and especially since 1960, studies of groundwater quality have expanded and in particular much emphasis has been placed on studying pollution and the movement of contaminants within groundwater systems. We have realized that just because groundwater is out of sight it is not totally out of harm's way, and that just because a polluted aquifer may not be as objectionable to our senses as a polluted river, that does not mean that groundwater pollution is less serious than pollution of surface water. We have also realized that contamination of groundwater is more difficult to detect and monitor, and the source more difficult to identify, than is the case with surface water. Perhaps most important, once an aquifer is polluted, the task of cleaning it up, even if possible, is enormous. Even more than in the case of pollution of surface water, the rule with groundwater pollution is that prevention is better than cure.

POLLUTANT MOVEMENT

The movement of a contaminant through a natural system is often referred to as **contaminant transport**, and the processes by which it takes place are called **transport mechanisms**. Several mechanisms are usually involved in the movement of a contaminant in groundwater. The simplest to envisage is probably the one that is usually called **advection**, although it is also referred to as convection. In this process the contaminant moves with the groundwater, usually in solution; it travels in the same direction and at the same speed as the groundwater.

We saw in Chapter 6 that the movement of groundwater through most aquifers is extremely slow, and that calculation of its velocity requires knowledge of the hydraulic gradient, the hydraulic conductivity and the dynamic porosity. All of these – especially the hydraulic conductivity – are properties that can vary greatly over short distances, and may not be known with any certainty. Thus the prediction of the movement of a contaminant, even when advection is the only process involved, may be difficult. As if that were not bad enough, the hydrogeologist trying to predict that movement has to contend with another problem – advection may not be the only process involved, so that the contaminant may not move with the same speed and direction as the water.

A process that plays a major part in pollutant movement is **dispersion**. If a small volume of a contaminant (or a chemical tracer representing it) is released instantaneously at a point in an aquifer, that tracer will not retain its initial volume and concentration. As a result of molecular diffusion and more importantly of mechanical dispersion, it will spread out both along (longitudinal dispersion) and perpendicular to (transverse dispersion) the flow direction, becoming diluted in the process (Figure 13.1a). This means that it will not arrive at a downstream sampling point at a single instant, but over a period of time that will increase with distance travelled.

Molecular **diffusion** occurs because a solute tends to move away from regions of greater concentration towards regions of lower concentration. If a small volume of a solute is introduced into a body of static water, molecules of the solute will move in all directions through the water away from their initial location. Diffusion is therefore a transport mechanism. The process is generally extremely slow, and it is only in rocks of very low permeability, in which groundwater flow speeds are very low, that diffusion is important in dispersing contaminants and tracers in groundwater. In most rocks **mechanical dispersion** is the main process that causes contaminants to spread out and be diluted, with molecular diffusion being much less important.

Transverse mechanical dispersion arises chiefly from the separation, recombination and tortuosity of the pore channels (Figure 13.1b), or of the fissures in a fissured aquifer. Longitudinal dispersion arises principally from the different speeds of groundwater in flow channels of different widths. This is particularly likely if the aquifer contains fissures in addition to intergranular porosity, so that preferential flow (p. 37) takes place, with some of the contaminant or tracer moving slowly through the intergranular pores and some being carried much more quickly through the more permeable fissures; flow through fissured formations is therefore especially

(a)

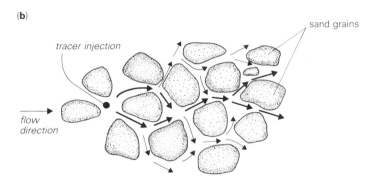

(b)

Figure 13.1 Dispersion. (a) Dispersion in a homogeneous isotropic aquifer. A fixed volume of tracer is released at the injection point A at time 0. At time t the tracer has reached B; after time t' it has reached C; after time t'' it has reached D. Note that dilution and dispersion increase with time and distance travelled, and that longitudinal dispersion is greater than transverse dispersion. (b) The process of mechanical dispersion in a sandstone.

dispersive. Attempts have been made to derive general relationships for the dispersion characteristics of aquifers, but it was soon recognized that minor variations in permeability and porosity within an aquifer can exert a major influence on the degree to which dispersion occurs at any specific locality.

A consequence of dispersion is that if a soluble contaminant is introduced into an aquifer at a point source – for example in the form of a leak of a chemical compound from a storage tank at factory A in Figure 13.2 – the compound will spread out to form a contaminant plume. If the contaminant has the same density as the groundwater it will follow the groundwater flow path, as in Figure 13.2; if it is more dense it will move at a different angle (Figure 13.3).

Dispersion complicates the task of predicting how a soluble substance will travel in groundwater. This is true even when the solute and the solid

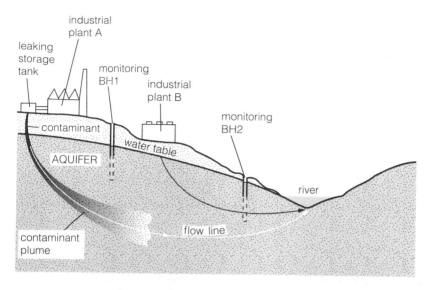

Figure 13.2 Contaminant transport. A soluble contaminant enters the recharge area of the aquifer beneath factory A. It moves along a groundwater flow line, undergoing dispersion as it is transported, to emerge eventually at the river. When the pollution is detected in the river, it is quite probable that factory B will be the prime suspect. By the time the pollution is detected, factory A may have closed and been demolished.

framework of the aquifer or aquitard have no effect on each other. An added complication arises if the solute is in some way affected by the framework, or affects the framework. Some substances may be **adsorbed** on the surfaces of mineral particles; many herbicides, for example, are adsorbed on the clay and humus particles of soils. Other substances may cause some particles, especially clays, to swell or shrink, altering the permeability of the formation and so changing the flow speed.

Two factors, perhaps above all others, have led us into a false sense of security over groundwater pollution. The first is the very slow speeds typical of groundwater movement. From the time a molecule of water enters the ground as recharge to the time it leaves the aquifer is rarely less than years, and may be centuries or millennia. The second factor, which often acts in conjunction with the first, is the enormous capacity of most aquifers to absorb and dilute pollutants. This arises from their size and the volumes of water they hold in storage, and from the effects of dispersion and adsorption. This means that a contaminant introduced into groundwater may be delayed so that it takes even longer than the water to emerge from the ground, and that when the pollutants do begin to appear, very often they are so dilute that they are not immediately detected.

With nothing to cause alarm, the activities causing the contamination may continue, until by the time the problem is discovered the total quantity

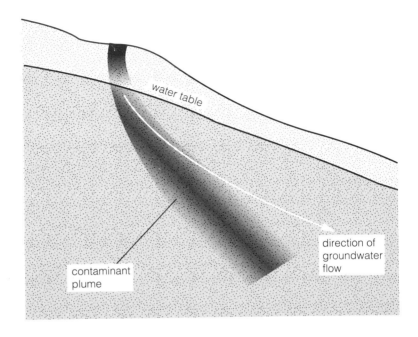

water table

direction of
groundwater
flow

contaminant
plume

Figure 13.3 Effect of contaminant density. If the soluble contaminant is much denser
than water, it will move downward at a steeper angle than the groundwater flow line.

of pollutant in and moving through the aquifer may be very large. In
combination with the increasing public concern in many countries about
drinking-water quality, the imposition of stricter standards, and the availability of increasingly sensitive detection equipment, this retention of
contaminants can lead to a situation where large volumes of groundwater
are suddenly discovered to be theoretically unfit for human consumption.

When talking of air pollution and the need to prevent it, people often
say that 'what goes up must come down'. We have to realize that the
nature of groundwater flow makes it almost inevitable that what goes
down must come up, however long it may take. And the problems of
removing a pollutant from an aquifer mean that it is much better to
protect groundwater than to try to clean it up after a contamination
incident.

WHO POLLUTES GROUNDWATER?

Few people set out deliberately to pollute groundwater, yet groundwater
becomes polluted, usually through accident, ignorance, expedience, or as

an incidental result of some other activity. The circumstances in which the contamination occurs – deliberate, accidental, or incidental – provide one possibility for classifying groundwater pollution, and the resulting classification could be potentially useful in deciding who should pay to clean up the aquifer. A problem with this approach, as we shall see, is that the majority of pollution arises incidentally.

We can classify groundwater pollution in other ways. We can, for example, split up pollution incidents according to the type of pollutant involved; is it microbial, or chemical, or radioactive? If a chemical, is it inorganic or organic? Another basis for classification is how the pollutant gets into the ground; as a first step, it is common to distinguish between pollution coming from single identifiable locations, which is referred to as **point-source pollution**, and that from more widespread sources, which is called **diffuse pollution**. Examples of point sources are leaking storage tanks, septic tanks, or waste-disposal sites. The most common diffuse pollutants are those applied to large areas of land during farming operations.

Not all pollution incidents fit neatly into these two categories. In a large industrial town there may be many individual sources of contamination. Contaminants from these points may move into an underlying aquifer, spreading out and polluting a large area. Such a collection of point sources is sometimes referred to as a **dispersed source**. Along a railway line, pesticide may be applied to combat weed growth; excess pesticide can travel through the track ballast and infiltrate to the water table. The same process may occur along highways, where other compounds such as de-icing salt may also enter the underlying groundwater. Do railways and highways represent lines of point sources, or very narrow diffuse sources? There is no real answer, and they are often referred to as **line sources**.

We should also consider whether the pollution continues over a long period – as perhaps from a leaking pipe – or whether it is an instantaneous occurrence, such as when a road tanker is damaged and spills its load.

In the following sections, we look at some of the substances and activities that do, or might, contaminate groundwater. This should not only give an insight into the ways in which groundwater can become polluted: it should also serve to show some of the many activities with which hydrogeologists are becoming involved.

SALINE INTRUSION

Saline intrusion is an example of incidental pollution that can occur in coastal aquifers. In a coastal aquifer, fresh water derived from recharge overlies saline water (Figure 13.4) in such a way that at the interface between them the pressure of fresh water usually exceeds the pressure of denser salt water, causing flow to occur from land to sea.

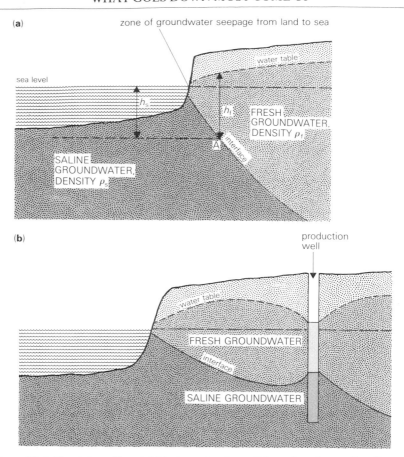

Figure 13.4 Coastal aquifers. (a) Under natural conditions there is usually a flow of groundwater from land to sea. The pressure of fresh water ($\rho_f g h_f$) at A equals or exceeds the pressure of saline water ($\rho_s g h_s$). (b) When pumping takes place, the lowering of the water table induces a corresponding rise of the interface; saline water migrates inland and may eventually reach the well.

If the natural condition is disturbed by pumping, the lowering of the water table results in a corresponding upconing of the interface, and salt water may be drawn into the well (Figure 13.4b). In practice the interface is not sharp but is affected by dispersion. Salt-water intrusion can also occur inland, particularly in arid regions, where fresh water overlies saline water. It is a problem that is receiving attention in many parts of the world, including Florida, The Netherlands and Israel.

On the south coast of England, near Brighton, a system of pumping from wells in the Chalk has been developed to limit the effects of saline intrusion. In winter, when there is a strong flow of groundwater into the sea, pumping takes place from wells near the coast to intercept this flow. In summer, these coastal wells are rested to reduce the danger of saline intrusion and pumping takes place from wells further inland, drawing on storage in the aquifer.

NITRATE

We saw in Chapter 11 that high levels of nitrate in water are deemed unacceptable by many regulatory authorities. In 'unpolluted' groundwater – by which I mean groundwater taken from an aquifer in a region where there is no agriculture, no significant human habitation and no significant content of oxides of nitrogen in the rainfall – the concentration of nitrate will usually be low, typically less than 5 mg/l.

For many decades, there have been individual wells in Britain and many other countries where nitrate levels in the groundwater have exceeded 50 mg/l or even 100 mg/l. Until the 1960s these could usually be explained as the result of continuous point-source contamination from badly-sited cesspits or septic tanks, leaking sewers, or farmyard manure. All of these are potentially important sources of nitrate, and are known still to be a major problem in many parts of the developing world.

By the late 1960s and early 1970s it became apparent that nitrate concentrations were increasing in all the unconfined aquifers in Britain, often to levels that exceeded 50 mg/l and occasionally approached the then maximum WHO limit of 100 mg/l. The effect was so widespread that it became clear that the source must be diffuse, with agriculture the prime suspect. During and after World War II there was a major effort in Britain to reduce dependence on imported food. Large areas of permanent pasture were ploughed up and turned over to arable farming, with cereal production a high priority. There was a change from the traditional crop rotation to cereal monoculture in those areas of Britain where soils and climate are particularly favourable for cereal growth. Many of these areas lie on the outcrops of aquifers in the eastern part of England – in particular on the Chalk and Jurassic limestones, which have light, well-drained soils, and generally low infiltration.

This intensive production of cereals, grown in the same fields for several years in succession, could be sustained only with the use of fertilizers – especially fertilizers providing nitrogen – and also of pesticides. In Britain, between 1940 and 1980, the average application of nitrogen in fertilizer to cereal crops increased from about 25 kg/hectare to about 150 kg/hectare, resulting in greatly increased yields. There has also been a move to more intensive livestock farming, making use of the higher densities of stocking that are possible on heavily fertilized grassland. This has led to a corresponding increase in the production of animal wastes.

The same trends in agriculture, and the same rising concentrations of nitrate in groundwater, were taking place in other countries – for example nitrate values were rising in the groundwater of the Chalk of the Paris Basin, with an estimated 300 drinking-water sources having to be closed as a result, and in Germany. More recently a survey by the US Environmental Protection Agency (EPA) found that nitrate was regarded by 41 states in the USA as representing a major threat to the quality of their groundwater, making it the most frequently reported contaminant in the survey.

The correlation between increased use of nitrogenous fertilizers and increasing concentrations of nitrate in groundwater seemed to be a fairly conclusive indication that the fertilizer was the source of the extra nitrate. It was assumed that some of the nitrate was being leached from the soil by infiltration and carried down to the water table. An obvious solution was for farmers to use less nitrate fertilizer, but the farming lobby in Europe was too powerful for such a course of action to be accepted without opposition.

Much more alarm was created in the early 1970s, after a study showed that relatively little tritium was present in groundwater in the Chalk; instead, most of the tritium that had entered in infiltration following atmospheric testing of thermonuclear weapons (Chapter 11) was still in the unsaturated zone. The tritium study suggested that water infiltrating the Chalk aquifer moves down through the unsaturated zone at a speed of less than a metre per year. As the unsaturated zone of the Chalk (Chapter 7) is frequently tens of metres thick, this implied that it could take decades for water to travel from the soil to the water table. This meant that the nitrate then affecting groundwater quality had been leached by recharge in the 1940s and 1950s, when rates of application of nitrate fertilizer were much lower. If this recharge was already causing problems, what would happen when the recharge from the 1960s and 1970s arrived at the water table, carrying its much greater contribution of nitrate?

To attempt to answer this question, and the more general one of whether anything could be done, major studies were started in Britain and other countries. These confirmed the presence of large quantities of nitrate in the unsaturated zone (Figure 13.5). They also revealed that the source of the increased nitrate is indeed agricultural, but that it is rather too simple to say that it arises directly as leachate from fertilizer.

Soil scientists point out that arable topsoil contains something like 5000 kg of nitrogen in each hectare, the nitrogen being bound up in organic matter. This compares with typical annual rates of application of fertilizer nitrogen to cereal crops of around 150 kg per hectare. The fertilizer is generally applied in the spring; unless heavy rain falls just after application, leaching some of the nitrate and carrying it away in runoff or infiltration to groundwater, most will be taken up by the crop. If all the crop were taken away at harvest, the nitrogen would go with it and there would be little nitrate to be leached into the water.

Instead, part of the crop remains behind as roots and stubble after the harvest. The harvest takes place at the end of summer, when the soil is still warm and just as the declining soil moisture deficits mean that moisture is becoming available – ideal conditions for soil bacteria that oxidize organic nitrogen to soluble nitrate. Ploughing, which traditionally takes place at this time, improves the entry of oxygen to the soil and so greatly helps the bacteria. If the soil is subsequently left bare over the winter then the soluble nitrate, with no plants to take it up, is leached out by the winter recharge.

Partial solutions to the problem are therefore to grow crops through the winter, so that as much as possible of the nitrate is taken up instead of being leached. This is done by planting cereals like wheat and barley in the autumn instead of the spring, or by planting a green fodder crop specifically to use

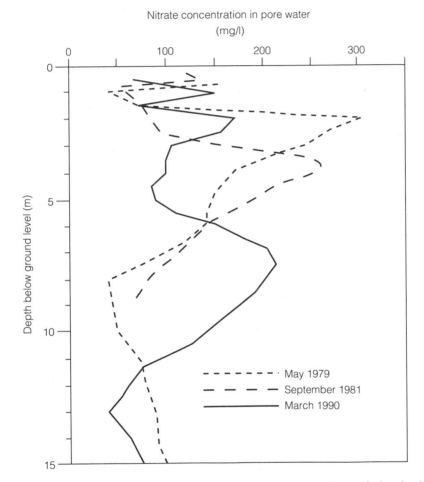

Figure 13.5 Nitrate concentrations in the unsaturated zone. The variation in the concentration of nitrate in pore water in the unsaturated zone of the Chalk beneath arable land in Cambridgeshire, England. The three profiles show how the concentration of nitrate in the recharge first increased dramatically (causing the peak at about 2.5 m depth in May 1979) and then decreased slightly. The peak value moves downward as more infiltration occurs and 'pushes' it downwards; it is flattened as a result of dispersion and possibly of localized denitrification. The increase in nitrate concentration may be due in part to high applications of nitrate fertilizer, combined with the effects of the 1975–76 drought. The subsequent reduction with time of nitrate at any particular depth probably results from the greater tendency to sow cereals in autumn, so that residual nitrate is taken up by plants rather than being leached from the soil by infiltration. (Based on data from the British Geological Survey.)

up the soil nitrogen. It also helps not to plough up a large proportion of the grassland in an area in any one year.

Many people point to the food surpluses being generated in Western Europe and North America using intensive agricultural practices and suggest that an obvious way to tackle the problem is to farm less intensively, perhaps switching to organic methods. Whatever the economic and political arguments for and against this course of action, it is worth pointing out that use of animal manures in place of fertilizer may lead to greater leaching of nitrate, partly because it is more difficult to control the precise amounts of nitrogen being applied. This is because manures are not as consistent as manufactured fertilizers, so that it is difficult to know the precise content of available nitrogen in a particular batch.

The European Community responded to the problem of nitrate in water by passing a Council Directive, published in December 1991, on the control of nitrate pollution from agricultural sources. The Directive (which under EC law must be implemented by member states) applies to both groundwater and surface waters. For groundwater, it requires member states to identify groundwater where the concentration of nitrate exceeds the limit of 50 mg/l specified in the Drinking Water Directive, or where this level is likely to be reached or exceeded unless action is taken. The member states had until the end of 1993 to designate the areas draining to these waters, and so contributing to the pollution, as 'vulnerable zones'. Member states then have until the end of 1995 to establish action programmes to reduce the pollution and until the end of 1999 to implement these programmes.

Britain responded to an earlier draft of the Nitrate Directive by instituting in 1991 an experimental scheme of Nitrate Sensitive Areas (NSAs). Ten areas, each representing the area of land contributing water to a public-supply borehole (a borehole catchment), were selected on a range of aquifers in different parts of the country. In these ten areas, farmers were invited to follow guidelines to reduce the amount of nitrate that would be leached from the soil. The practices to be followed included not leaving land bare in winter, and restricting the amounts and times of application of manures and fertilizers.

Eighty-seven per cent of eligible land was entered, with those farmers joining the scheme receiving compensation for the loss in production estimated to result from following the guidelines. Additional payments were available to farmers who converted arable land to low-intensity ('Premium Scheme') grassland; this accounted for 14% of the total area of the NSAs.

Initial findings based on samples of water from the unsaturated zone in these areas suggest that the scheme has indeed reduced the nitrate concentration in recharge water moving down towards the water table, but the concentration still exceeds 50 mg/l. Also, the results may have been distorted by the drought of 1988–92, which may have left excess nitrate in the soil waiting to be leached. Although the guidelines resulted in greater use of nitrogen by the crops, and therefore less nitrogen being leached to groundwater, one of the measures needed to achieve the total reduction was the export of manure from livestock units in the NSAs. This means that nitrogen is being exported from a borehole catchment, and may lead to higher rates of application elsewhere.

A second, larger, group of NSAs was being defined in 1994 around public water-supply wells in the Chalk, Permo-Triassic sandstones and Jurassic limestones. As with the original scheme, membership is voluntary and farmers who join will receive compensation. When the Nitrate Vulnerable Zones (NVZs) required by the EC Directive are defined, however, there will be no compensation for the farmers affected. Most European countries seem to be moving towards the idea of making the NVZs large – perhaps covering the whole of an aquifer outcrop or the whole country – but making the restrictions imposed on farmers within the zones fairly gentle. Britain, in contrast, appears to be opting for smaller zones around water-supply sources, but imposing strict limits on the application of nitrogen within the zones. This could be seen as unfair to a small proportion of farmers who happen to be farming near public-supply wells.

Although some people argue that much of the nitrate problem has been caused by over-intensive agriculture, it has to be said on behalf of the farmers that initially they were merely responding to government and popular pressure to grow more food, particularly during World War II. It also seems unfair that a farmer whose land has been farmed for generations suddenly finds himself in difficulty because a water undertaking chose, relatively recently, to sink a well on or near his land. Nitrate pollution is therefore a prime example of incidental pollution – the people accused of causing it were carrying out a perfectly lawful occupation, generally following the best practices of their profession, unaware until recently that they could be causing a problem.

PESTICIDES

Pesticide is a general term for all substances used to kill pests, whether the pests are animals or plants. It therefore includes insecticides, herbicides (often referred to in the domestic role as weedkillers), and fungicides.

Before World War II, the pesticides in use were generally inorganic or natural substances. They included pyrethrum, compounds of arsenic, copper, lead and zinc, and cyanides. Some had been used for centuries. Many were toxic to humans, and some resulted in deaths of workers, but the very fact that they were so obviously toxic meant that adequate precautions were usually taken.

By the time of World War II, a combination of factors led to the introduction of synthetic pesticides. To begin with, there was the need. Troops in the forests and jungles of south-east Asia were at risk from malarial mosquitoes and other insect-borne diseases; refugees crowded together were susceptible to lice carrying typhus and other diseases, which had killed millions earlier in the century. Because of demand, cost and wartime problems, traditional insecticides were not available in sufficient quantities. There was the possibility of producing synthetic substitutes, and the capacity to do it; studies of substances that might be used in chemical warfare revealed that some of them were deadly to insects, and the expanding chemical industry was able to produce them in large quantities and relatively cheaply.

DDT (dichloro-diphenyl-trichloroethane) was first synthesized in 1874, but it was not until 1939 that its properties as an insecticide were recognized. DDT is one of a group of substances called organochlorines – compounds that contain carbon and chlorine. A second group, the organophosphates (compounds of carbon and phosphorus) includes malathion and many other insecticides.

In the 1950s and 1960s, these synthetic organic pesticides were regarded as among the greatest achievements of science. The growing world population needed more food, which could only be supplied by more intensive use of land. The high yields that have been achieved since the 1950s have been made possible by the use of fertilizers and pesticides.

Medically, DDT was successful in killing lice and so preventing the spread of typhus; for a time it looked as though it would also help to eradicate malaria by killing the mosquitoes that carry the malarial parasite. The organochlorine pesticides are very persistent. At first, this was seen as an advantage; DDT crystals, for example, could be sprayed onto the walls of houses and would stay there, poisoning mosquitoes, for months. Refugees and soldiers were dusted with it and suffered no ill effects, because in powder form it is not readily absorbed through the skin.

Then problems began to appear. One was that strains of pests began to develop that were resistant to the new pesticides. Others were that pesticides were seen to affect beneficial organisms, and their safety began to be questioned. It was discovered that combinations of some of them are much more toxic than the toxicities of the individual combinations would suggest, posing unexpected risks to people applying different substances.

Pesticide residues began to be found in invertebrates, fish and birds in concentrations that seemed high compared with the rates of application. This happens because DDT and other organochlorine pesticides are soluble in fat. Taken in the diet in minute concentrations, they accumulate in the bodies of invertebrates such as earthworms, so that their bodies may contain concentrations many times higher than those ingested; when the invertebrates are eaten by higher animals, the higher concentration received by those animals is itself concentrated again in their fatty tissue. Because of the persistence of the organochlorine pesticides, they began to accumulate in the food chain, and pesticides began to be found in habitats where no pesticide had been applied.

Some of these concerns began to be expressed in the late 1950s, but it was the publication in 1962 of Rachel Carson's book *Silent Spring* that brought them to the attention of the world at large. Carson correctly predicted that once pesticides or their residues entered groundwater they would be difficult to remove and would spread to surface waters. In reality it was not until the early 1980s that pesticides were detected in groundwater, but they have now been found in many aquifers, albeit usually at very low concentrations. To detect such low concentrations, sophisticated analytical equipment and techniques and relatively large samples of water are needed. Care has to be taken when collecting, storing and transporting samples so that the minute amounts of pesticide or pesticide residue are not 'lost', perhaps by being adsorbed on the surfaces of sampling equipment or containers.

Since Carson's book was published many new types of pesticide have appeared. In 1985, in the United States alone, an estimated 500 000 tonnes of pesticides were used at a cost of more than $4500 million, a figure estimated to grow by about 5% per year. Tens of thousands of different chemicals or mixtures of chemicals are now used as pesticides. They include the triazine group of herbicides, and the carbamate group, which includes both insecticides and herbicides. In Britain, the greatest use of pesticides is of herbicides to control weeds, particularly on land used to grow cereals.

When a pesticide is applied to land, whatever the 'target' organism, much of the applied pesticide finds its way onto the surface of the ground and so, usually, into the soil. Here, many pesticides are adsorbed, particularly onto organic material (humus), while some are also strongly adsorbed onto the surfaces of clay minerals. Generally speaking, the larger the pesticide molecules the more likely they are to be adsorbed by organic matter.

Pesticides strongly held in the soil by adsorption are unlikely to be leached into soil water and so into aquifer recharge. This is an important point, because many of the new pesticides are much more soluble in water than were the organochlorines; some have solubilities of hundreds or thousands of milligrams/litre, millions of times greater than concentrations permitted in drinking water. When held in the soil or in plant tissue the pesticides may degrade, though in some cases the products of degradation may themselves be as toxic as, or more toxic than, the original pesticide.

Pesticides are broken down in various ways, but degradation by soil micro-organisms is the single most important process. Organochlorines break down very slowly in most soils. Organophosphates include some acutely toxic chemicals, but various micro-organisms cause them to degrade much more quickly than organochlorines. The same is true of many of the most widely used herbicides, with the exception of the triazines. This last group includes atrazine and simazine, herbicides which are widely used in the United States to control weeds among maize crops.

In Britain there is limited use of atrazine and simazine in horticulture except in orchards. It was therefore something of a surprise when the first surveys for pesticides in groundwater revealed these substances to be present, and they have subsequently been widely found in groundwater, albeit at low concentrations. The most likely reason is that triazines were used as general defoliants on railways and on highway verges.

In these environments several factors combine to increase the likelihood of pesticide reaching the water table. To begin with, they are used as total defoliants, so they are applied at high rates. Secondly, spraying represents a disruption to normal operation. On railways special trains have to be operated to carry out the spraying. On major roads trucks or teams of workers have to move slowly along the road, causing potential delays to traffic. The authorities do not want these disruptions to occur more times than is absolutely necessary, so there is an obvious temptation to apply a generous dose of herbicide to ensure that a repeat application will not be needed. Much of the herbicide will not be taken up by the plants. For safety reasons, highways and railways are well drained, and where they cross aquifers that drainage is often led into the ground, bypassing the soil and the organic material and clay minerals that would adsorb the herbicide. Because of the increasing occurrence of atrazine and simazine in groundwater, their non-agricultural use in Britain has been banned since August 1993. The concern now is that the herbicides that replace them may be more mobile, toxic or persistent than triazines, and that the water industry may be less well placed to detect them or remove them from water supplies.

The occurrence of triazines in groundwater serves to highlight a more general concern. This is that many agricultural herbicides are used on the outcrops of aquifers, which are often among the most intensively-farmed land in Britain. On sandstone aquifers the soils are often sandy and permeable, with little in the way of organic material or clay minerals to provide adsorption capacity for herbicides. On Chalk and other limestones, the soils are usually thin; they contain some clay but again relatively little organic material. There may thus be little to adsorb and retain pesticides in the soil zones of these aquifers. Once pesticides have passed below the soil, the concentrations of micro-organisms decrease markedly (Chapter 11), so that, even for those substances that are readily degradable, there is less likelihood of them being broken down.

The nitrate issue has shown us that it may take a long time for a substance that enters the unsaturated zone to reach the water table of an aquifer and become detected in a borehole. It is clear that monitoring of the unsaturated zone must be our first line of defence against diffuse pollution of groundwater by agricultural processes and chemicals. Given that pesticides are much more toxic than nitrate, it is perhaps as well for us that widespread application of fertilizer to, and release of nitrate from, farmland occurred before the widespread use of agricultural pesticides, and provided us with a warning of what can happen.

WASTE DISPOSAL BY LANDFILL

Landfills were formerly referred to more mundanely as rubbish dumps. Landfill is the practice of disposing of waste materials by placing them in an excavation and – usually – covering them with soil or other non-waste material. The excavation can be above or below the water table, in aquifers or in poorly-permeable formations, and can be lined or unlined. The wastes disposed of in landfills can vary from relatively harmless inert material, such as the rubble produced when a building is demolished, through domestic and garden waste, waste from commercial operations (which may include such diverse material as waste paper and offal from slaughterhouses), to toxic wastes produced by industry.

The practice of landfill grew up to replace previous disposal methods which consisted simply of burning rubbish or dumping it on the ground surface. Early landfill sites were usually abandoned quarries, pits, or open-cast mines, which were filled with rubbish, often to level the ground prior to its being restored for some other use; such sites developed around many industrial towns in Europe and North America. Waste was simply dumped into the excavations.

Before World War II, most industries in Britain disposed of waste by burning or burying it on or near the operational site – often leaving a legacy of **contaminated land** that has still to be tackled. Most homes had open fires; these chiefly burned coal, but provided a means for disposing of small quantities of cardboard, paper and other flammable waste. Relatively few products were prepackaged, little canned food was available, and bottles were usually returned and recycled. These facts were reflected in the domestic waste produced. In 1935, in terms of weight, at least half the content of the average British dustbin was ash; paper and cardboard made up about 15% and vegetable and other putrescible material about the same. There was little glass, textile or metal, and no plastic.

By 1990, ash and other fine material accounted for less than 15%, paper and cardboard for 30% and vegetable and putrescible matter for nearly 25%. The proportion of metals had risen to 9%, glass to 8% and plastic to 7%. The total quantities of waste have also increased, so that there is now a shortage of available sites. Instead of mineral extraction providing a valuable product and coincidentally leaving a hole which can be filled with rubbish, the situation has been reached in Britain where the hole can be more valuable, as a waste-disposal facility, than the minerals extracted from it.

In a landfill containing putrescible waste – which includes the **sanitary landfills** used by local authorities for the disposal of domestic and garden rubbish – decomposition begins when aerobic bacteria get to work. The action of these bacteria consumes the oxygen that was incorporated in the waste and produces carbon dioxide in much the same way as the working of a garden compost heap; this first stage of decomposition generally takes from a few weeks to a few months (Figure 13.6).

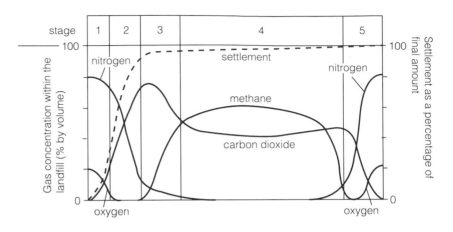

Figure 13.6 Stages of decomposition in a landfill.

When the oxygen has been consumed, anaerobic bacteria take over and produce more carbon dioxide and fatty acids. Methane also begins to be produced. In stage three, the fatty acids are themselves broken down to produce more methane. Stage four is effectively an equilibrium phase in the life of the landfill. The characteristic feature of this stage – which may continue for tens of years – is the generation of relatively constant levels of methane and carbon dioxide, the two main components of landfill gas. The fatty acids are broken down by methane-producing bacteria at about the same rate as they are formed. The landfill is effectively functioning as a bioreactor – a system in which micro-organisms reproduce and break down the waste material. Populations of different bacteria are existing together within the landfill, with the products of one group providing the energy source for another.

During stages one and two of the life of the landfill, settlement occurs as the lower layers of waste are compressed beneath overlying material. As decomposition of the waste gets under way in stages two to four, solid material is effectively turned into soluble compounds and gases. As rainwater or subsurface water infiltrates these landfills, it can dissolve and react with some waste components and products of decomposition to form leachate. Leakage or removal of leachate and gases effectively removes mass from the site, leading to further settlement. Eventually a stable framework of non-degradable material is developed, and settlement effectively ceases even though decomposition is still taking place.

The leachate formed in a landfill is potentially a serious source of contamination of groundwater. Even with landfills intended for the disposal of inert or relatively innocuous waste, there have been instances of unscrupulous operators or individuals disposing of harmful or toxic waste, often at night, so it is rarely safe to assume that a landfill will not produce leachate that is potentially contaminating.

In the United States, groundwater contamination from landfills became a serious issue in the late 1970s, with the discovery of groundwater contamination from the Love Canal waste site in New York.

In Britain, public concern about the possible danger to groundwater from landfills arose in the early 1970s, when drums of waste containing cyanide from electroplating industries were discovered in landfills in the Permo-Triassic sandstones of the Midlands. These sandstones are an important source of public water supply in the area. Legislation was enacted in the form of the Control of Pollution Act (Part I) (COPA I) of 1974, which dealt with the disposal of waste to land.

The parts of COPA associated with the licensing and operation of landfill sites have now been superseded by the Environmental Protection Act (1990). This Act is administered in England and Wales by the Waste Regulation Authorities (WRAs)*, who must consult the National Rivers Authority (NRA) over the granting of licences for handling and disposing of waste; if the NRA believes that a WRA is proposing to issue a licence to an operation that does not meet requirements to protect water, it can object to the appropriate Secretary of State. The Environmental Protection Act provides additional powers, and requires that sites are operated by competent persons, that operating companies have adequate provision for long-term management and remediation, and that operating licences cannot be surrendered (i.e. responsibility relinquished) until the site no longer represents a threat to the environment.

The migration of gas from unlined landfill sites can pose a significant hazard. A high proportion of methane is usually present. Methane is explosive at concentrations between 5 and 15% by volume in air. Because methane is less dense than air it may move through permeable strata, beneath less permeable layers, until it has the opportunity to escape. The less permeable layers may include waterlogged soil, made ground, pavement or building foundation slabs, and the opportunity of escape may be provided by an unlined basement or the entry of a service duct into a building.

Landfill gas also contains carbon dioxide. This is denser than air and so can accumulate in depressions and cause suffocation.

There were several cases of problems caused by landfill gas in Britain in the 1980s; the sudden occurrence of these problems is attributed to the increased amount of putrescible material in domestic waste, and to changes in practice with material being placed more quickly and at higher density. Probably the most widely-publicized example occurred at Loscoe in Derbyshire in March 1986, when gas migrating from a landfill collected beneath the suspended floor of a bungalow, from where it is believed it

*In 1996, the NRA and WRAs are due to merge with Her Majesty's Inspectorate of Pollution (HMIP) to form the Environment Agency.

seeped into the rooms above and was ignited by the pilot flame of the central-heating boiler early one morning. The bungalow was completely destroyed, and the three occupants (who were in bed at the time) were badly injured but fortunately survived.

Immediately after the implementation of COPA I, the favoured practice in Britain was the use of so-called 'dilute and disperse' landfills (commonly referred to as attenuation landfills in North America). The principles of this type of landfill are the attenuation of contaminants by physical, chemical and biological processes in the unsaturated zone, and the dilution of leachate in groundwater. In this type of landfill there is little or no attempt to contain leachate; it is allowed to enter groundwater and mix freely with it. The logic is that after this attenuation and dilution, the concentrations of any toxic substances will be too low to pose a threat to health or the environment.

In 1980 the European Community published Directive 80/68/EEC, on the protection of groundwater from pollution caused by certain dangerous substances (commonly referred to as the 'Groundwater Directive'). This Directive classifies the dangerous substances into two lists, List I and List II, depending on toxicity. List I includes substances such as mercury and cadmium and their compounds, cyanides, hydrocarbons and organo-chlorines. List II includes metals such as lead, arsenic and their compounds, biocides not covered by List I, ammonia and nitrites. The Directive prohibits the direct or indirect discharge to groundwater of substances in List I, and limits discharges of substances in List II. Because many of the listed substances are contained in landfill leachate, pressure mounted for an end to 'dilute and disperse' landfills.

In the United States, the Resource Conservation and Recovery Act (RCRA) of 1976, and subsequent amendments, the Hazardous and Solid Waste Amendments (HSWA) of 1984, regulate the disposal of waste. They are administered by the EPA, and generally require high standards of containment and monitoring to protect groundwater.

As a result of the need to protect groundwater, to prevent gas migration, and to comply with legislation, there has been a move away from the practices of 'dilute and disperse' towards a principle of containment and control. Good practice now requires that wastes that may produce polluting leachates should be disposed of in landfills designed to contain the products of decomposition. In these landfills, the production of leachate is also reduced by reducing infiltration of rainfall at the surface.

To achieve containment the landfill must be sited in or lined with material of low permeability. The liner may be made of mineral material such as clay or a mixture of soil and bentonite, or of synthetic material (usually referred to as a **geomembrane**) such as high-density polyethylene (HDPE) sheeting 2.5 mm thick, or a combination of the two. In Britain, the minimum requirement is usually a clay layer 1 m thick, compacted so that the hydraulic conductivity does not exceed 10^{-9} m/s (about 10^{-4} m/day).

Modern containment landfills are engineering structures. Any construction is likely to be imperfect; good engineering design recognizes this, and makes allowances for the fact that a structure may not be built precisely to the engineer's specification. A synthetic liner will be made in a factory, and will have a uniform low permeability. When it is put in place, however, it may be torn or damaged, leaving a few discrete weak points where leachate can escape. Also, the material is of finite width, so many widths have to be welded together to cover the base and sides of a typical landfill; each weld is a potential line of weakness.

In contrast, a clay liner is largely self-repairing – if a hole is punched in it, the clay will usually 'squeeze' to seal the hole again. However, the clay will be placed by people using earth movers and rollers. They may inadvertently leave an area of higher permeability – perhaps where the clay has incorporated some sand or silt, or not been properly rolled. Clearly, neither clay nor geomembrane alone provides an ideal barrier to leachate movement; a combination of the two is much better, because the head of leachate applied to the clay as a result of a hole in the membrane will be applied only over a very small area. Such a combination is called a composite liner.

A typical composite liner (Figure 13.7) might therefore consist of a layer of clay, 1 m thick, placed in and compacted in thin layers, followed by an HDPE membrane placed firmly against the clay to prevent movement of fluid between the clay and the membrane. Above this will be a layer of fine sand or geotextile (an artificial permeable textile) to protect the membrane, followed by a drainage layer to collect leachate and keep the head across the liner to a minimum. Drainage pipes in the layer usually lead the leachate to collection sumps. A layer of geotextile protects the drainage layer, and the waste is placed on top of the geotextile.

Leachate generation is reduced by reducing the entry of rainwater. This is done by operating the landfill as a series of cells, each separated from its neighbour by a wall or bund of soil or compacted waste. As each cell is filled it is capped with clay or other material of low permeability to reduce infiltration, covered with soil, and planted with vegetation to help reduce infiltration. The restored surface is usually domed to encourage runoff and to make some allowance for future settlement. In practice, it is difficult to allow for settlement.

Produced leachate is collected and, if necessary, treated and disposed of. This prevents leachate from overflowing from the landfill, and also reduces the head of leachate available to create a hydraulic gradient and cause leachate to flow through the base or sides of the landfill. The gas produced is also collected, and either used as a fuel or flared to dispose of it.

Monitoring boreholes are installed around the landfill, so that any leachate that does escape and contaminate the groundwater can be detected. Migration of landfill gas can also be identified and rectified as early as possible.

Reducing infiltration by the use of a capping layer helps to keep down

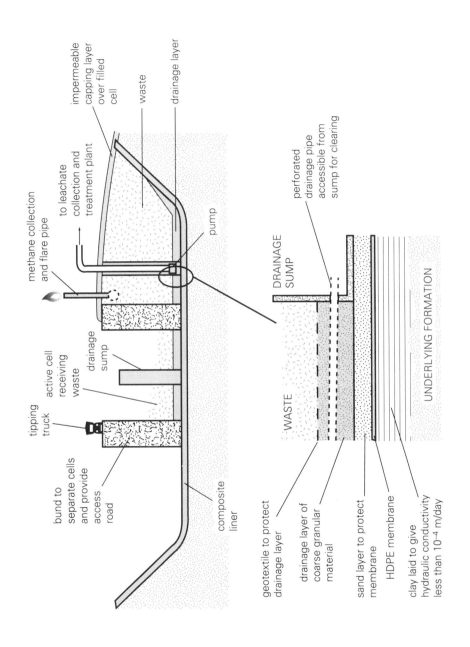

Figure 13.7 Section through a landfill with enlargement of liner.

the production of leachate, but may also mean that bacteria have insufficient moisture to work effectively to break down the waste. This could be serious, for it would mean that the waste may not be broken down for hundreds of years; our direct knowledge of the life expectancy of lining systems extends for only about 30 years. Under the Environmental Protection Act, the holder of the waste-management licence for a landfill in Britain remains responsible for it until the landfill is stable and the WRA allows surrender of the licence. This could mean the landfill operator or licence holder being required to maintain leachate and gas control facilities and monitoring systems for perhaps several hundred years. A way of reducing this time is to recirculate the leachate within the landfill, adjusting the pH and nutrient levels, to encourage the bacteria in their task.

NUCLEAR WASTE

If disposal of domestic waste in sanitary landfills presents hydrogeologists with a technical problem, it is as nothing compared with the problems posed by the need to dispose of waste from the nuclear industry. This waste is normally grouped into three categories: low level, intermediate level and high level.

Low-level radioactive waste consists of material that is not radioactive in the sense that most people would use the word. It consists largely of material that has been used in laboratories, hospitals or power stations, in environments where radiation may be present. It may therefore have become contaminated, and must be treated as such. It includes such diverse items as laboratory gloves and coats, and steel and concrete from decommissioned reactor buildings (not reactor cores).

Intermediate-level waste consists of materials that emit radiation at sufficient intensity that they require shielding, but that do not generate significant amounts of heat. It includes the cladding from fuel elements, and material from decommissioned reactors, such as graphite from cores and steel from pressure vessels and coolant pipes.

Smallest in volume, but of most concern to the public, high-level waste consists mainly of used fuel from nuclear reactors, or the wastes generated by the reprocessing of that fuel. This waste contains a number of radioactive isotopes with half-lives extending to millions of years. In addition to being highly radioactive, a characteristic of high-level waste is that it emits heat.

High-level waste is normally stored underwater in special bays at the sites where it is generated. This method of storage removes the heat generated by radioactive decay. Although it would theoretically be feasible to store all nuclear waste in buildings in nuclear facilities, this option is considered unacceptable. Some of the waste will be a danger to life for so long into the future that we would, in essence, be leaving our descendants to pay much of the cost of energy that we have used. The most dangerous wastes would have to be guarded against terrorist attack and protected against environmental changes that we can only speculate upon. The buildings would certainly have to be reconstructed many times. We also cannot be sure that future civilizations will have the technology to deal with the problem.

For these reasons it has been decided that high-level nuclear waste must not just be stored: it must be disposed of, in such a way that it will pose neither an expensive problem nor a threat to the health and environment of succeeding generations. This is usually taken to mean that the waste must be isolated from the human environment for at least 100 000 years. Although various options have been considered, the general consensus is that the most suitable method of disposal is to incorporate the waste in some relatively inert material such as borosilicate glass and bury it deep below ground – the so-called geological option.

The main drawback to this approach is groundwater, which is the medium that could transport radioactive material back to the human environment. The principle on which disposal is to be based is that of using multiple barriers, which will act together to limit the rate at which groundwater can leach radio-isotopes from the waste and carry them to the surface.

The barriers include the borosilicate glass, the metal containers holding it, concrete or similar materials surrounding the containers in the underground cavity created to hold them, and the surrounding rock (Figure 13.8). The man-made materials are termed **engineered barriers**, and form the **near field** of the repository, and the surrounding rock is the **natural barrier** and forms the **far field**.

When this option was first proposed, it might have seemed relatively straightforward. Geologists knew of many rock types in many countries that are 'impermeable', in the sense that they would not be expected to yield water to wells. Unfortunately, when these rocks began to be investigated in detail, problems soon became apparent. One problem is that although a rock may have too low a permeability to be relied upon as a source of water, its permeability may be very significant when tens of thousands of years are available for water movement. Another problem is that the heat given out by radioactive waste may have a detrimental effect on some potential host rocks such as clay or rock salt. A third problem is that in rocks that are almost impermeable, the limited pathways that do exist for water flow – such as fissures in granites or metamorphic rocks – are often unevenly distributed, making it difficult to predict how flow will take place if it does occur. Studies have revealed that such pathways exist in many rocks, and we often do not know about them only because we have never studied those rocks in detail.

In the 1970s and 1980s many countries with nuclear programmes began seeking sites for waste repositories. The potential 'host' rocks investigated included clays (for example in Belgium), salt (Germany and Denmark), volcanic tuffs (USA), basalts (USA) and granites and metamorphic rocks in several countries including Sweden, Canada, Switzerland and the United Kingdom. At a disused iron ore mine in metamorphic rocks at Stripa, in

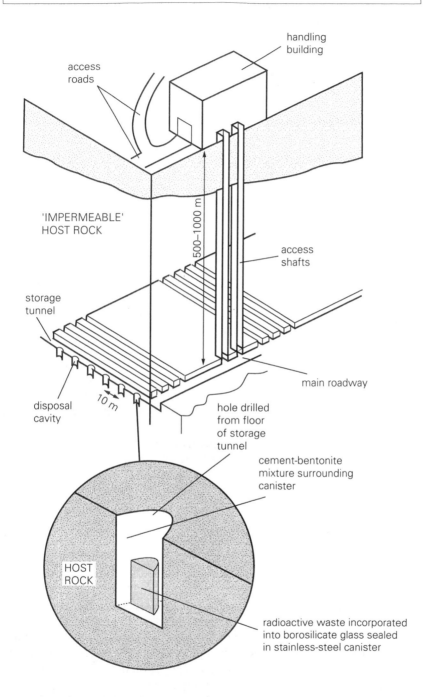

handling
building

access
roads

'IMPERMEABLE'
HOST ROCK

500–1000 m

access
shafts

storage
tunnel

disposal
cavity

10 m

main roadway

hole drilled
from floor
of storage
tunnel

cement-bentonite
mixture surrounding
canister

HOST
ROCK

radioactive waste incorporated
into borosilicate glass sealed
in stainless-steel canister

Figure 13.8 A repository for nuclear waste. A diagrammatic representation of a possible layout of a repository for the disposal of nuclear waste. The inset shows the engineered barriers.

Sweden, scientists from several nations cooperated to develop techniques, including advanced geophysics and complex packer testing, suitable for evaluating potential repository sites. Canada carried out evaluations at several sites on the Canadian Shield, and constructed an Underground Research Laboratory (URL) in a granite near Pinawa in Manitoba.

These studies are very different from hydrogeology in the sense that the word was used before 1970. They have brought together people from many disciplines, including engineers, physicists, mathematicians, chemists and microbiologists, as well as geologists and geophysicists. These teams have access to budgets and resources that most conventional hydrogeologists can only dream about, and they have made enormous progress in understanding the behaviour of fluids in fractured rocks and in other poorly-permeable materials. Have they solved the problems? Some of the scientists and engineers would argue that we could now dispose of nuclear waste in such a way that the risk to the general population would be far less than the risks that the population faces – or chooses to face – from many other activities. Others would say that the hazards involved are potentially so great that the problems have not been solved satisfactorily, and perhaps never will be. The final decision will have to be a political one, but a political decision informed as well as it can be by the scientific evidence.

In Britain the political decision-making so far is not encouraging. In the 1970s the United Kingdom, in common with other countries with major nuclear-power or nuclear-weapons programmes, began research into suitable sites for the geological disposal of high-level waste. These investigations concentrated on areas of igneous and metamorphic rock.

In 1981, largely because of political fears about the public opposition that was emerging in all the areas being investigated, the UK Government announced that it was postponing the search for a disposal facility for high-level waste. The reason given for this was that the high-level waste would anyway need to be stored on the surface for about 50 years to allow much of the activity to decay before eventual disposal, so there was no pressing need for research.

The Government announced a plan to find a suitable site for disposal of low-level waste by shallow burial, with the aim of using deep burial for intermediate-level waste. Low-level waste has traditionally been disposed of by shallow burial at Drigg in Cumbria, near the Sellafield nuclear site, initially in unlined trenches that were operated very like a conventional landfill.

Shallow disposal sites were sought in poorly-permeable clay formations in eastern England, with four sites being selected for detailed study. Again there was major opposition from local residents, and in 1987, with a general election looming, the Government again announced a change of plan and said that low-level as well as intermediate waste would be disposed of by deep burial. Currently, investigations for this repository are taking place at Sellafield, where an underground testing laboratory is proposed. The host rock here would be the Borrowdale Volcanic Group, the rock formation that forms England's highest mountains in the Lake District a few kilometres to the east.

HYDROCARBONS

Some hydrocarbons are soluble in water, but the majority that cause contamination problems in aquifers are almost insoluble. These relatively insoluble hydrocarbons are often referred to by hydrogeologists as **non-aqueous phase liquids**, or NAPLs for short. They are subdivided into those that are denser than water, which are called dense NAPLs or DNAPLs, and those that are less dense than water, the light NAPLs (LNAPLs).

Light NAPLs consist mainly of petroleum products, such as petrol, diesel fuel, jet fuel, lubricants and chemical feedstocks. They can enter the ground as a result of spillages, accidental rupture of vehicle fuel tanks or of transport vehicles, and leaks from storage tanks and pipelines (especially those sited below ground). In 1989 it was estimated that there were between 2.8 million and 5 million underground storage tanks in the United States containing liquids, other than wastes, that could be hazardous to groundwater; the largest group is those containing hydrocarbons, such as at filling stations. Of the total, it was estimated that as many as 450 000 were leaking. The US EPA set in train regulations under HSWA (even though the hydrocarbons are not wastes) to improve the situation. These regulations in essence demand that underground storage tanks are designed and monitored to ensure integrity, very much along the lines of containment landfills.

The most important group of dense NAPLs is the chlorinated hydrocarbons. These include compounds such as carbon tetrachloride, trichloroethene and perchloroethene. These and similar compounds are used in metalworking industries as degreasants, in the electronics industry, in the tanning of animal skins, and as dry-cleaning solvents. Chlorinated hydrocarbon is another name for organochlorine compounds; we tend to use the former term generally, and the latter term more specifically for pesticides. Pesticide manufacture, wood preservatives and transformer oils can also be sources of DNAPLs. The guide level laid down in the European Drinking Water Directive for chlorinated hydrocarbons is 1 microgram/litre, which corresponds to about one part in 1 000 000 000.

Non-aqueous phase liquids are almost invariably non-wetting liquids relative to water, but wetting liquids relative to air. Put simply, this means that if NAPL comes into contact with dry rock (one with largely air-filled pores), the NAPL will be drawn into the pores by the capillary forces. If the NAPL comes into contact with a water-saturated rock, the NAPL will not enter the pores unless it is forced in under pressure. This has important implications for the behaviour of NAPLs in the unsaturated and saturated zones.

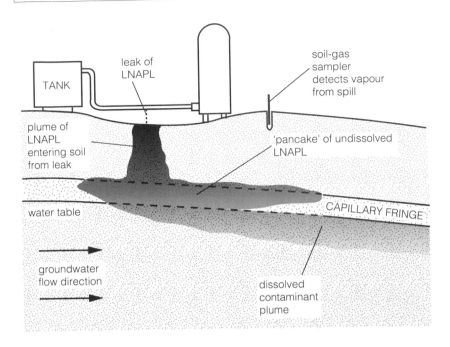

Figure 13.9 Penetration of LNAPL into an aquifer. The light liquid flows vertically through the unsaturated zone until it reaches the capillary fringe and the water table. There it spreads out as a thin layer 'floating' on the water. Soluble fractions dissolve in and move with the flowing groundwater.

When an NAPL comes into contact with the soil or unsaturated zone it will enter the pore space when there is sufficient head acting on it to force it through the connecting pore throats and so make it displace the air or water that was previously occupying the pores. This will be easy in a material with large pores that are mainly filled with air, such as a gravel or coarse sand, and even easier in air-filled fissures; it will be relatively easy in water-filled fissures or large pores, but will be much more difficult in a fine material such as clay or unfissured chalk that is largely saturated with water.

Downward migration of the NAPL through the unsaturated zone will continue if the source is large enough and continues to supply the necessary pressure for entry into the pores. This is most likely to occur where a large head of NAPL is available – perhaps because a pool of spilled liquid has formed on the soil surface, or where the liquid is leaking from a buried pipe under pressure, or from an underground or partly buried storage tank containing a large head of the liquid. If the supply and head are not available, the NAPL will become discontinuous and be trapped as globules or films in pores and fine fissures. Dense NAPLs move downward under gravity more readily than LNAPLs, not only because of their greater density but because they are usually much less viscous.

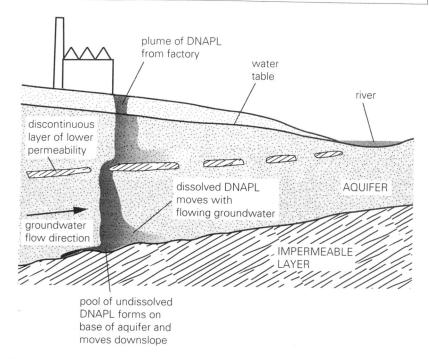

plume of DNAPL
from factory

water
table

river

discontinuous
layer of lower
permeability

dissolved DNAPL
moves with
flowing groundwater

AQUIFER

groundwater
flow direction

IMPERMEABLE
LAYER

pool of undissolved
DNAPL forms on
base of aquifer and
moves downslope

Figure 13.10 Penetration of DNAPL into an aquifer. The dense liquid penetrates quickly and deeply into the aquifer, being stopped or delayed by layers of lower permeability. At the impermeable base of the aquifer, the DNAPL spreads along the junction, moving down the slope under the influence of gravity. It begins to dissolve into and move with the flowing groundwater.

On reaching the top of the capillary fringe, LNAPL will spread out and flow along the capillary fringe. If the downward flow of LNAPL from the surface is fast enough, it will depress the capillary fringe and spread along the water table, forming a 'pancake' (Figure 13.9). As the water table rises and falls, perhaps seasonally or as a result of pumping, the LNAPL will be distributed through a greater thickness of the aquifer.

Dense NAPL will not usually be stopped at the water table unless the quantity is so small that insufficient head develops to force it into the saturated pores. It will continue downwards until it meets a layer of material of sufficiently low permeability that it cannot penetrate any further downwards. It will then move down the slope of this layer under gravity (Figure 13.10), collecting in any small hollows on the surface.

If NAPLs were completely insoluble in water, that might be the end of the matter, with the light liquids floating on the water table and the dense liquids resting on the base of the aquifer. Although such a situation might be intuitively undesirable, the liquids would do little harm unless a borehole penetrated the concentration; even then, with the light liquids floating on the water surface above the pump suction, and the dense liquids resting on the base of the aquifer, there would be little chance of them being drawn up and entering a water supply.

In reality, these conditions do not persist. The NAPLs are not completely insoluble. Fuels and lubricants are mixtures of many hydrocarbons, which have greater or lesser solubilities, all of them generally low. One exception is MTBE (methyl tertiary butyl ether), the compound that has replaced lead as an 'anti-knock' agent in unleaded petrol; it is significantly more soluble than most of the components of fuel. Chlorinated solvents are also slightly soluble. In all cases the maximum solubilities are usually measured in milligrams/litre or tens of milligrams/litre, but these have to be compared with the permitted levels in drinking water, which are generally lower by three or more orders of magnitude. As the NAPL dissolves in the ground-water, it generates a plume of contamination moving in the direction of groundwater flow (Figures 13.9 and 13.10). This plume moves with the groundwater, regardless of whether the NAPL is light or dense. In the case of DNAPL, we may therefore have a situation in which the undissolved DNAPL is moving one way under gravity, but the slowly dissolving DNAPL is moving in the opposite direction under the influence of the hydraulic gradient (Figure 13.10).

Many of the compounds – especially the chlorinated hydrocarbons – are also volatile. They vaporize and mix with the air in the unsaturated zone, and the vapour travels through the air-filled pores by diffusion. It can then re-dissolve in the pellicular water, and be carried down to the water table, thereby spreading the contamination to other parts of the aquifer. On the positive side, the presence of the vapour in the soil gas above a contaminant plume can provide a means for determining the extent of the plume using a **soil-gas survey**. Samples of soil gas are collected from a metre or so below the soil surface, using a simple probe, and analysed using an instrument called a gas chromatograph; this technique is much cheaper than drilling boreholes to the water table and taking water samples for analysis.

Light NAPLs are generally amenable to being broken down by bacteria – they are said to be **biodegradable**. Partly for this reason, and partly because they do not penetrate far into the saturated zone, they are generally perceived as less of a threat to groundwater than DNAPLs, although LNAPLs enter aquifers in much larger total quantities. DNAPLs are generally resistant to biodegradation; they break down only slowly by the chemical process of hydrolysis, and can remain in an aquifer, dissolving slowly into the ground-water or vaporizing into the unsaturated zone, for a very long time.

CLEANING UP THE PROBLEMS

Removing contaminants from aquifers is not easy. For example, once an NAPL has entered the pore space of an aquifer, withdrawing it is rather like trying to pump oil from an oilfield. Any petroleum geologist can tell us that it is difficult to extract more than about 30% of the oil in place in an oil reservoir by pumping; some oil comes out relatively easily, but the remainder separates into individual globules or films, and is difficult to withdraw. The same is true of NAPLs in aquifers. A particular problem in removing LNAPLs such as fuels is in skimming them off from the water table.

Some NAPLs can be removed by pumping the water and NAPL mixture to the surface and passing it through a device called a separator to separate the water and the NAPL. This method is most commonly used for fuels and lubricants. Volatile liquids, such as some of the chlorinated solvents, can be removed from the pumped water by passing the water through an air-stripper, which blows a current of air through a stream of falling water and evaporates the solvent. Both of these belong to a group of methods referred to as **pump and treat**; in either case the treated water is disposed of by re-injecting it into the ground or putting it into a stream.

A variation of air-stripping is to carry it out in the ground, when it is known as **soil vapour extraction**. Boreholes drilled through the contaminated zone to just above the water table are connected to a vacuum system which draws air through the soil into the boreholes. As it passes through the soil the air removes volatile contaminants. The recovered air is then treated to remove water and contaminant vapour before being discharged to the atmosphere. The process can be combined with a technique called **sparging**, in which air is compressed and injected through other boreholes below the water table; as the air migrates upwards, volatile substances are carried with it to the unsaturated zone, whence they can be removed by soil vapour extraction. Sparging can also be used to provide oxygen to encourage bacterial growth as an aid to *in-situ* bioremediation (Chapter 11).

These and other new technologies are being developed rapidly to help clean up the contaminated sites and the associated contaminated groundwater left at thousands of industrial sites around the world. If the problems confronting the people trying to restore these sites were only the technical ones, the task would be daunting enough. Unfortunately, other factors have to be considered.

Imagine the following situation: a borehole is used to abstract water from a sandstone aquifer to supply the local area. One day, routine analysis shows that the water contains traces of a chlorinated hydrocarbon solvent. The sampling and analysis are repeated, and the contamination is verified. The borehole is taken out of supply, and water has to be diverted from an adjacent supply area.

Within 2 km of the borehole is a modern industrial estate. Enquiries reveal that several units on the estate use chlorinated hydrocarbon solvents, or have done so within the last few years, but all deny that any has been lost or been disposed of illegally. Nevertheless the water company drills monitoring boreholes between its contaminated well and the industrial estate. Minute traces of solvent are detected in some of them, but in concentrations far lower than those found in the production well.

A soil-gas survey is carried out, with some difficulty in this partly urban area. It indicates the presence of a contaminant plume, coming not from the industrial estate but from a residential area. A more detailed survey and further trial drilling confirms that the plume appears to be emanating from the gardens of some modern and expensive houses. It transpires that the houses have been built on a site that was formerly occupied, in part, by a factory that machined metal components. During and after World War II, the factory used chlorinated solvents to clean the metal parts, and disposed of dirty solvent by tipping it into a hole on the site. The houses have been built on contaminated land.

By now, the water company and the agency responsible for monitoring water quality have spent tens – possibly hundreds – of thousands of pounds on drilling, testing and analysis, not to mention the cost of rearranging the water supply. They are seeking redress. In Britain, and in most other countries, the principle in a case of this type can be summarized as 'the polluter pays' – the person or persons who caused the pollution must pay to clean it up, and compensate for any damage caused. Under the most recent British legislation they must do this even if they did not intend to cause the pollution; the Water Resources Act of 1991 makes it an offence 'to cause or knowingly permit' any poisonous, noxious or polluting matter or any solid waste matter to enter any 'controlled waters', which include groundwater and water in the unsaturated zone.

But who is the polluter? The owners of the houses claim, with justification, that they had no knowledge of the contamination. The builder of the houses purchased the site from a developer, who had demolished the derelict factory and levelled the site some years previously. The company that owned the factory has ceased trading – it may have gone into liquidation, or the man who ran it as a family business has died. Even if the owner can be traced he can claim quite truthfully that, at the time he was operating, he was doing nothing illegal, and could have had no idea that his action would one day cause a threat to a public supply which might not even have existed at the time. It is an issue of principle in most countries that legislation should not be retrospective.

Although the case just outlined is imaginary, it contains many elements that have arisen in real life. One real case that caused much interest in the water industry concerned a water supply near Cambridge, in eastern England. The Cambridge Water Company discovered perchloroethene in groundwater being pumped from the Chalk aquifer at Sawston. Detailed investigations over several years detected a plume of perchloroethene dissolved in groundwater flowing towards the production borehole from a tannery, where perchloroethene had been used, about 2 km away. It is assumed that solvent had leaked through cracks in concrete floors and yards, and from pits used to dispose of waste fat from animal hides, into underlying gravels and so into the Chalk beneath. It had then moved down under gravity, with small amounts ponding on less permeable layers, until it collected in a 'pool' at the base of the aquifer. The perchloroethene then dissolved slowly in the groundwater, forming the plume of contamination.

Cambridge Water Company brought a civil action against the owners of the tannery, claiming compensation of more than £1 million, the amount it had spent installing a new borehole to provide a replacement supply.

The case ultimately went to the House of Lords, who reversed a previous decision of the Court of Appeal and found that the water company was not entitled to the damages sought. The basis for the decision was that in establishing civil liability in these circumstances it is necessary to show that harm could be foreseen in the actions of the defendant. At the time the leaks occurred, there was no way that the owners of the tannery could have foreseen that the perchloroethene would render water unfit for human consumption on the basis of a European Directive that had not then been passed.

If such leaks occurred now, it would be much more difficult for a defendant to claim that harm could not be foreseen. It would still be difficult for a plaintiff to establish the source of the groundwater pollution; at Sawston, the polluted groundwater had moved relatively quickly – within a few years – from the tannery to the borehole because it had travelled through fissures in the Chalk. Even so, it required a major study, funded largely by central government for research purposes, to establish the source of pollution. In many aquifers, the solvent might not have been detected at the borehole for decades. In English civil law, for a plaintiff to win a case he or she has to establish the fault of the defendant on the balance of probability. In criminal law, the fault has to be established beyond all reasonable doubt. If it is difficult for one company to win a case against another in civil law, it is that much more difficult for a regulatory authority such as the NRA to bring a successful prosecution in the criminal courts against a polluter of groundwater.

Given all the problems of determining the source of contamination, finding the individual or company responsible, and convincing the courts, who – in an example like the first one above – is going to clean up the aquifer, restore the water supply, and compensate the water company for the expense? For some years, British hydrogeologists and engineers looked enviously at the United States, where in 1980 the Comprehensive Environmental Response, Compensation and Liability Act (CERCLA) was passed, along with accompanying legislation called the Superfund Tax Act. The latter legislation placed a tax on oil and some hazardous substances. The money was 'banked' to create a fund – the **Superfund** – for cleaning up contaminated sites.

CERCLA created a system for identifying sites that pose a threat to health or the environment, for cleaning them up, and for recovering the costs of clean up from the 'polluter'. The US EPA identifies the site, investigates the problem and authorizes the remediation, with the cost being

met initially from the Superfund. Those believed responsible for the problem are then sued by the US Government to recover the costs, and the
Superfund is reimbursed. The primary, and very laudable, aim of CERCLA
was to identify and remedy releases of contaminants into groundwater.

The legislation has not been an unqualified success, although it has
created significant employment opportunities for environmental consultants (including hydrogeologists) and for lawyers. Given the difficulties
outlined above in restoring some contaminated sites, it may be that some
of the objectives set for some sites were too ambitious, with large sums of
money being spent in trying to achieve standards that were not practically
attainable.

Another problem has been the effect on members of the public
suddenly finding that a site in their neighbourhood – perhaps adjacent to
their homes – has been designated a 'Superfund' contaminated site. The
designation can have an immediate and dramatic effect on the value of
property, so that homeowners, who have lived for years near what is
regarded as an untidy but harmless former industrial site, suddenly face
double trauma of finding that the area is considered hazardous and that
they cannot easily leave it because their homes have become almost impossible to sell. People who had unknowingly purchased property containing
hazardous substances discovered that they had inherited the liability for
remediation along with the title deeds.

A modification to CERCLA removes the liability for clean up from
property owners if they can show that they had no reason to know of the
contamination. In practice, to claim this immunity, the property owner
needs to have an environmental survey of the property carried out before
purchasing it, to show that at the time of purchase the property had a clean
bill of health. This too has created work for environmental consultants. A
proposal to reform CERCLA to remove and simplify some of the
problematic sections of the legislation and reduce the money being spent
by litigation was defeated in the Senate in Autumn 1994, shortly before the
US mid-term elections.

Partly for fear of antagonizing property owners, the British Government
has backed down from its earlier plan of creating a register of contaminated
land. It is now working towards a system of requiring such land to be
restored to a level that takes into account its actual or intended use. The
problem with this approach is that it may not necessarily take into account
the potential for harm to be done to groundwater if, for example, the water
table rises and leaches toxic solutes from the site.

GROUNDWATER PROTECTION

Given the difficulties, technical, legal, social and financial, of restoring
contaminated groundwater, it is obviously preferable to ensure that

contamination does not occur. This brings us to the concept of groundwater protection.

There are two aspects to the protection of groundwater. The first is that of protecting the groundwater **resource**. This recognizes that all groundwater is a valuable resource, whether it is used directly by abstracting it from wells or springs, or indirectly by allowing it to flow to surface watercourses from which water is subsequently taken; or whether it is simply allowed to form or augment surface waters for the benefit of the environment. A meeting of European Community environment ministers at The Hague in November 1991 effectively endorsed this principle. The problem in applying it, as we have seen, is that most acts that cause groundwater pollution are not deliberate, but arise from some other activity. However much we wish to protect groundwater, it is not practical to prevent all of those other activities; what we have to do is minimize the chance that they will have a detrimental effect on groundwater.

We therefore have to distinguish the concepts of **hazard** and **risk**. Put simply, a hazard is something that could be dangerous or harmful: a stick of explosive or a tank filled with a solution containing arsenic are hazards. Given that we sometimes need sticks of explosive or tanks of arsenic, we cannot eliminate the hazards that these constitute: what we have to do is try to minimize the risk of them causing injury. We would not, for example, place the tank of arsenic next to a children's playground, or store explosives in a wooden hut in the middle of a housing estate in an area prone to bush fires.

In protecting groundwater, it is necessary to identify substances and activities that represent a hazard to groundwater and try to minimize the risk of them causing contamination. An important consideration here is the way that the groundwater occurs. Groundwater in an aquifer confined beneath a thick layer of relatively impermeable clay will be much less vulnerable to pollution than groundwater in fissured chalk at outcrop. Many activities that could be undertaken above the clay without posing a significant risk to the groundwater would be regarded as posing an unacceptable risk if carried out on the Chalk; the operation of an unlined landfill would be an obvious example. Therefore if a landfill is to be allowed on the Chalk at all, something must be done to reduce the risk to groundwater – by ensuring a very high standard of containment, and perhaps placing some restrictions on the waste that can be deposited there.

We therefore have a basis on which to construct a policy to protect aquifers. We have to consider the nature of the *hazard*; next we have to look at the importance of the aquifer, the nature of the flow mechanism within it, and the degree to which its groundwater has natural protection – overlying impermeable beds, for example, or thick soils that will offer the opportunity for adsorption and microbial degradation of some contaminants. Depending on this degree of natural protection – or, looked at from the opposite point of view, on the aquifer's **vulnerability** to contamination

– we can then decide which activities should be allowed in particular areas, and what special requirements should be imposed to reduce the risk from other activities.

In imposing these requirements, regulatory agencies do not always have a free hand. A company that has been operating a factory on the outcrop of an aquifer for many years may be using outmoded techniques that pose a risk to the groundwater; if told that it must invest large sums of money to reduce the risk, it may well say that it cannot afford to do so, and will have to close, throwing hundreds of people out of work. In general in cases like this a principle is emerging called BATNEEC – the best available technology not entailing excessive cost. This means that the company must do what it reasonably can to improve the situation, without being forced to do something so expensive as to put its operation in jeopardy. Cynics claim that we are likely to find that what is actually installed is CATNIP – the cheapest available technology not involving prosecution.

In the real world there are always going to be shortcomings with aquifer protection – the balance between risk to groundwater and other factors such as cost will inevitably sometimes go wrong. Accepting this, we have to try to minimize the risk to groundwater in the areas where it is arguably most important – the places where groundwater is abstracted for public supply. These public-supply springs and wells are referred to in Europe as **public sources**, and so this brings us to the second aspect of groundwater protection: that of protecting not the whole resource, but of protecting the *sources*.

Because the areas around the sources are smaller than the outcrop areas of aquifers, more stringent constraints can be imposed on the activities within the **source protection zones** (SPZs). The concept and implementation of source protection zones are well established in some European countries. In England and Wales provision for SPZs was made in Part II of COPA, but the idea was never implemented. The NRA is currently in the process, with the aid of consultants, of defining SPZs around the major groundwater sources used for public supply.

How do we set about defining an SPZ? The first step is to recognize that water abstracted from a well has to come from somewhere. When pumping begins from a well, the cone of depression will expand as water is taken from storage; if there is no recharge to the aquifer coming from elsewhere, water must continue to be taken from storage and this expansion must continue. Usually there is a source of water. In an unconfined aquifer, the source will be recharge derived from the excess of rainfall over evapotranspiration; the cone of depression will expand until the area of land it influences is large enough to capture an amount of recharge which on average is equal to the rate at which water is abstracted from the well (Figure 13.11). Thus if the annual average recharge is r (measured in metres), and the annual abstraction from the well is Q (m^3), then the area of the cone of depression must be A m^2, where

$$Q = Ar.$$

If the aquifer is confined, then the cone of depression must expand until it either meets an outcrop and captures recharge in the unconfined part of the aquifer as described above, or until it captures leakage through the confining beds, or captures a combination of recharge and leakage.

If the potentiometric surface of the aquifer were initially horizontal, we could therefore define a protection zone around a borehole simply as the area of the cone of depression (Figure 13.11). This would be a total protection zone – all groundwater and any pollutants inside it would eventually reach the well, and no groundwater or pollutant outside it would do so; the cone of depression would be the same as the **borehole capture zone**.

Usually, however, potentiometric surfaces are not horizontal, and so the cone of depression and the capture zone are not the same. Further complications arise when – as is usual – the aquifer is not homogeneous and isotropic, so that the cone of depression is not circular, and when aquifer boundaries (Chapter 10) cause further distortion. Then computer models normally have to be used to define the capture zone, often with all the potential problems entailed in using models when insufficient data are available.

Do we need to protect the whole of the capture zone, or just part of it? We have to accept that any water, and any pollutant, that enters the aquifer inside the capture zone will eventually reach the well *unless it is destroyed on the way*. Pathogenic bacteria, for example, may die long before they can reach the well; other contaminants may decompose or be broken down by micro-organisms before they can get there, provided that the journey takes long enough.

It is therefore customary to define subsidiary protection zones that are smaller than the total capture zone of the well. These zones are usually defined on the basis of the time it would take a water molecule to reach the well from the outer edge of the zone, with a minimum distance specified. In making the calculations of travel time it is usual to ignore the time taken from the ground surface to the water table. This provides an additional element of safety, as does any adsorption or retardation of the pollutant relative to a water molecule.

Commonly used SPZs are a 'courtyard' zone immediately around the well, in which nothing is allowed other than activities associated with water production; an inner zone corresponding to a travel time of 50 days or a radius of 50 m, whichever is larger; an outer zone corresponding to a travel time of 400 days; and the total capture zone or catchment to the source. The logic is that most pathogenic organisms will die in an aquifer in less than 50 days, and many chemical pollutants will decompose in less than 400 days.

Within the 50-day zone we therefore need to restrict any process – such as the use of septic tanks or cess pits – that could allow pathogenic

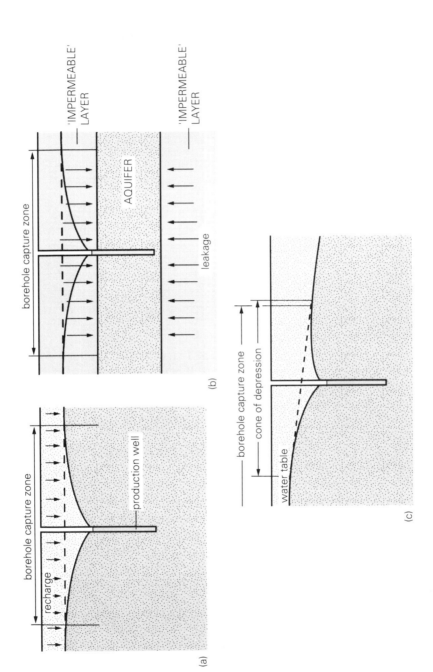

Figure 13.11 Cone of depression and capture zone. If the water table or potentiometric surface is initially horizontal. then the borehole capture zone is the same as the cone of depression. (a) In an unconfined aquifer, the cone of depression expands until it captures sufficient recharge to supply the amount being abstracted from the well. (b) In a confined aquifer, the cone expands until it captures sufficient leakage from or through the 'impermeable' beds above and below. (c) If the potentiometric surface or water table is sloping, the capture zone will differ from the cone of depression. The capture zone will not extend as far as the cone does downgradient; upgradient it will extend to the groundwater divide.

organisms to enter the aquifer, but we can be more lenient outside this zone. In the same way, more restrictions need to be imposed on the use of some chemicals within the 400-day zone than outside it. However, some of these figures are beginning to be questioned, and there are arguments to err on the side of caution.

To calculate these travel times we need knowledge of hydraulic conductivities and dynamic porosities. In the case of fissured aquifers, the flow speeds may be so high that a water molecule can travel from the source catchment boundary to the well in less than 50 days, so that the whole catchment needs the same protection as the inner zone.

In densely populated areas with low annual recharge, underlain by highly transmissive fissured aquifers, the protection zones will be large and will frequently meet up. In these circumstances we are faced in effect with the need to give most of the aquifer the same degree of protection as that laid down for the inner protection zone. In some cases this will simply not be compatible with the other activities that have to go on in such an area if normal life is to continue, and compromise will be necessary, but at least the compromise will be informed.

Defining protection zones is not easy. Defending them when they are challenged by developers who wish to build factories, garages or chemical works within them will also not be easy. The point in their favour is that restoring polluted aquifers is even more difficult.

SELECTED REFERENCES

Bedient, P.B., H.S. Rifai and C.J. Newell 1994. *Ground water contamination; transport and remediation*. Englewood Cliffs, NJ: PTR Prentice-Hall.

Brady, N.C. 1990. *The nature and properties of soils*, 10th edn. New York: Macmillan Publishing Company.

Carson, R. 1962. *Silent Spring*. New York: Houghton Mifflin. (1963, London: Hamish Hamilton.)

Domenico, P.A. and F.W. Schwartz 1990. *Physical and chemical hydrogeology*. New York: Wiley.

Foster, S.S.D., P.J. Chilton and M.E. Stuart 1991. Mechanisms of groundwater pollution by pesticides. *Journal of the Institution of Water and Environmental Management* **5**, 186–93.

Freeze, R.A. and J.A. Cherry 1979. *Groundwater*. Englewood Cliffs, NJ: Prentice-Hall (see especially Ch. 9).

Marco, G.J., R.M. Hollingworth and W. Durham (eds) 1987. *Silent Spring revisited*. Washington, DC: American Chemical Society.

Ministry of Agriculture, Fisheries and Food (MAFF) 1993. *Solving the nitrate problem*. London: MAFF.

National Rivers Authority 1992. *Policy and practice for the protection of groundwater*. Bristol: National Rivers Authority.

Parker, J.M., C.P. Young and P.J. Chilton 1991. Rural and agricultural pollution of groundwater. In: *Applied groundwater hydrology* (eds R.A. Downing and W.B. Wilkinson). Oxford: Clarendon Press.

Todd, D.K. 1980. *Groundwater hydrology*, 2nd edn. New York: Wiley. (See especially Chs 8 and 14.)

Williams, G.M., C.P. Young and H.D. Robinson 1991. Landfill disposal of wastes. In: *Applied groundwater hydrology* (eds R.A. Downing and W.B. Wilkinson). Oxford: Clarendon Press.

In addition, the Construction Industry Research and Information Association (CIRIA) is in the process of publishing a 12-volume series on the remediation of contaminated land.

Epilogue

14

The science of hydrogeology developed because people needed to get water supplies from the ground. It has been said that there are four stages in an orderly process of obtaining and using groundwater for supply: exploration; development (or exploitation); monitoring (or inventory); and management. At first, in the Old World, development and exploration went together and little science was needed; the benefit of experience, passed from one generation to the next, was available to tell people where and where not to dig wells. But as industry and agriculture began to demand ever larger supplies, local experience was not always adequate; and as demand arose in newly settled areas like the American Mid-West and the African colonies, the experience was often non-existent. Then exploration began to go systematically ahead of exploitation, and the need developed for investigation and scientific guidance. This led to the work of people like Darton and Meinzer of the United States Geological Survey (USGS), and of Frank Dixey in the Colonial Surveys in Africa.

To begin with, the emphasis was often simply on finding suitable aquifers. Later, the realization that resources were finite and the evidence of declining heads in some places (as in the Dakotas and the London Basin) caused more attention to be given to understanding aquifers, measuring their properties and estimating their resources. In the United States, Hubbert, and Theis and other scientists of the USGS, made major contributions during this period, which can be thought of as the beginning of monitoring. In Britain, the realization that excessive abstraction in some areas was causing problems led to the introduction of legislation to control abstraction by a system of licensing. This was the beginning of the era of management.

Since World War II, as we have seen, groundwater studies have broadened to include studies of water quality and of pollution and contaminant transport. Monitoring now involves quality as well as quantity, and models are increasingly used to predict the movement of contaminants and the results of the mixing of waters of differing quality.

Hydrogeologists can, and do, take part in exploring for and understanding the origins of hydrocarbon accumulations and mineral deposits. They are increasingly involved in environmental issues, including assessment of the impacts of developments such as mineral extraction and waste disposal. Many hydrogeologists now regard the development and management of groundwater resources – once the reason for the existence of their science – as a relatively small part of hydrogeology.

Yet about one thousand million people on this planet lack the basic amenity of a safe and reliable supply of drinking water (Figure 14.1) – and that is despite the efforts devoted to the problem in the International Drinking Water Supply and Sanitation Decade from 1980 to 1990. During

Figure 14.1 Water problems in the developing world. Children in a Bangladesh village take advantage of water flowing in an irrigation channel from a deep borehole. Despite the efforts of governments, charities and aid agencies, many villagers still have to walk hundreds of metres to obtain drinking water.

that decade, many communities were provided with such a supply, but the births of many more millions of people meant that the number without such provision fell only slightly.

This is a sobering fact, because the Earth's population in 1994 was estimated to be about 5700 million; by 2050 it is estimated that it will probably be around 9000 million and could be as high as 12 000 million. By then, there will have been a significant shift in population distribution. It is estimated that for the first time in human history, more than half of the population will live in cities. And this great influx of people will not be living in leafy suburbs, well-served by water mains and sewerage; it is probable that a large proportion of the extra people on Earth will be living in shanty towns and squatter camps.

The problems in 2050, as now, will only in part be technical. War, politics and greed will all play their part in making life wretched for many of our fellow citizens, just as they have done this century and for centuries past. There will be enough water for everyone, though it may not be readily available where it is most needed. To monitor it, conserve it (whether in aquifers or reservoirs), protect it from pollution, and distribute it equitably, are problems that need technical solutions. Making the necessary resources available is a political and economic problem.

The history of human development shows that we can always find the necessary technical solutions, but that we frequently lack the resolve to use them in time for the benefit of all our fellow citizens. As far as water supply is concerned we have, or will develop, the technical knowledge to solve the problems of 2050. Will we have the political wisdom and the will to apply it?

Index

Page numbers appearing in **bold** indicate major entries or definitions; numbers in *italic* refer to figures.

INDEX